Man and the Living World

Also by Karl von Frisch: THE DANCING BEES

Man
and the
Living World

Karl von Frisch

Translated by Elsa B. Lowenstein

A HARVEST BOOK

A HELEN AND KURT WOLFF BOOK

HARCOURT, BRACE & WORLD, INC.

NEW YORK

Illustrations by Henri A. Fluchere

First published in Germany under the title *Du und das Leben:
Eine moderne Biologie für jedermann*

Library of Congress Catalog Card Number: 62-16734

Printed in the United States of America

Living Nature offers us
Unbounded scope for wonder,
With open mind and open eyes
In humbleness to ponder.

CONTENTS

side issue. Knights in armor. Many tiny animals build mountains.

Walking and motoring. Why we get hot when we are running and why we get out of breath. Movements in animals
and plants.

Fuel and building materials. The basic foodstuffs. Vitamins.
We are vegetarians, after all. How plants feed. Why arable
land has to fertilized.

What happens when we digest? Praised be the art of cooking.
The other side of the story. The appendix, a nuisance and a
help. The lost glass beads and other strange stories.

Heart and blood vessels. Why we do not bleed to death from
a pinprick. A drop of blood under the microscope. The cir-
culating blood and its function. The police. Immunity and immunization. Heartless animals, and such as have a heart in
every leg.

The lungs and their ventilation. Why Man cannot breathe
under water and fish suffocate in the air. Life without oxygen.

Why we need a winter coat and why the sparrows and lizards
do not. The regulation of the body temperature.

A journey of exploration of our own body. There are more
things between heaven and earth . . .

Master noses. The sense of smell in insects — a strange love
story. Tasting with the toes.

Organs of equilibrium. Mrs. Cricket is called to the telephone.
We thread our way through a labyrinth. Can fish hear?

From the eye spot to the lens eye. The world topsy-turvy.

Man and the Living World

I

LIFE, DEATH, AND IMMORTALITY

1. *About the Span of Life*

The word "biology" is taken from the Greek and means the science of life. What is life? This is sooner asked than answered.

Just as we often appreciate a thing we own most fully after we have lost it, so the death of our canary, the beloved pet of the family, makes us recognize what made it alive and gives us understanding of the characteristics of life itself. Only the day before the bird had hopped about, chirped, and picked at its food, and we could feel the warmth of life as it settled trustingly on our hand.

Now it lies stiff and cold. But does this contrast really signify the difference between all forms of life and death? The apple tree in the garden does not sing and does not move about, it is not warm, we do not see it eat, and yet it is alive.

"Well, that is a plant," you will say. But this is not a full answer. The sponge we use for washing is the dead skeleton of an animal that lives at the bottom of the sea and which is brought to the surface by the sponge fishers. The fresh sponge is different from the dead skeleton we buy in the store; it has a slimy covering which represents the living substance. However, though it is alive, it is in no way more attractive than the toilet sponge in our bathroom. We do not see it eat and it does not move about. It does not even squirm when we cut it to pieces, and yet it is a living animal. As we can see, our question about the characteristics of life is not so easy to answer. Let us therefore put the question aside for a moment and return to our canary, which, unquestionably, is not alive any more but dead.

This death saddens everybody, and the youngest in particular cannot understand it at all. The father tries to explain that the bird was old, and that all living beings, whether animal or plant, have to die too when they have grown old. This is how we react: when we are faced repeatedly with the same happening, we come to take it for granted. But do we really have to take it for granted that all living things must die?

How long does life last?

If one were to consult a big tome in which all the known facts about the duration of life had been carefully collected, one would find that death is indeed the common lot of animals as well as of plants.

It is also obvious that no age is spared. A young blade of grass may be crushed, a newly hatched bird may fall prey to the cat. But let us leave aside such accidental death. Even when they are protected from all perils and placed in the most favorable environment, there comes a time for Man, animal, and plant when they grow weak with age and die a natural death. A human being may live up to a hundred years or so. Most do not reach this noble age; some animals die very much earlier. Some insects are known to live just a few hours as winged adults. However, these insects happen to have spent their youth, lasting from several months up to several years, as aquatic larvae. Some worms die of old age a few days after hatching, some flies after a few weeks, some beetles after a few months, other insects live for years. Snails live six or seven years, dogs up to ten or twelve years, chickens fifteen to twenty years, and pigeons forty to fifty years. Each kind of creature has a "natural" span of life. Man himself can be outlived by any parrot or elephant, which may live to the age of one hundred to one hundred and fifty years, or by giant turtles, which are said to live between two hundred and three hundred years and thus seem to hold the record for longevity among animals. A still greater diversity is to be found in the plant kingdom, where some trees, such as our yews, the cedars of Lebanon, and the famous redwood trees (Sequoia) of California, grow perhaps to be a thousand or so years old. During their life span not only generations of human beings but successive periods of human culture have come and gone, until at last these giants too grow old.

Nature allows no exceptions, so how could we hope to see our canary escape the common lot? But again: Do we have to take it for granted that all things living must die? Is this invariably so? To get to the bottom of this question we have to look at living things much more closely.

2. *The Building Blocks of Plant and Animal Bodies*

We need a magnifying glass

The inquiring scientist likes to study his objects under a maximum amount of magnification. He looks through the tube of a microscope, both ends of which carry specially ground lenses, magnifying objects up to 1,500 times or so.

Until recently, this seemed to be the limit of possible magnification. It was thought that under the microscope light cannot produce a distinct image of a structure that is of a similar or smaller dimension than the light waves themselves. Recently, however, human inventiveness presented us with the electron microscope in which electron beams take over where the light rays let us down. The beam employed in the electron microscope is directed not by glass lenses but by electromagnetic fields. In this way we get an image which, though not immediately visible to our eyes, can be photographed or projected onto a fluorescent screen. The old barriers are down. Within a few years we have thus learned to magnify objects a thousand times smaller than those that used to escape our view. Now we photograph even molecules, and they are very small units of matter indeed.

A new world reveals itself

The electron microscope is a very complicated apparatus, so far used in laboratories only. But even the ordinary microscope not only offers the naturalist enlarged details of visible objects but reveals to him an otherwise invisible world. The unaided eye cannot see a living organism that is about 1/250 inch long. But magnify it a thousand times and it will appear as long as our index finger and quite a monster.

The microscope not only reveals organisms which because of their smallness we had not noticed before, but permits us to observe the otherwise invisible details of larger objects.

If, for instance, a bit of onion is cut with a sharp razor and is magnified 50 to 100 times, it can be seen to consist of many small units, known as cells (Fig. 1). They are a common feature of all plants, and each cell is separated from its neighbor by a firm cell wall. It is therefore not surprising that after the invention of the microscope the cells were first discovered in plants and got their name because of their shape, which often resembles the cells of the honeycomb in a beehive.

What does a piece of our own or of an animal's body look like under the microscope? If we happen to slice off a piece of skin from our finger tip with the bread knife and, having an inquiring mind, put it under the microscope, what we see will be rather

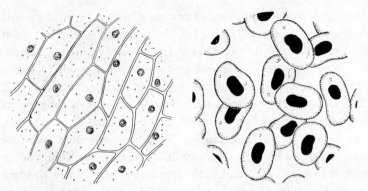

Fig. 1. Cells under the microscope, plant and animal. Left, cells of the onion; right, red blood cells of a frog.

disappointing. We will see nothing, because the portion of the skin is too thick and not transparent enough. To cut a thin enough slice from this soft and elastic skin, a special method is needed which will be described later. Then the skin will show structures similar to those found in the onion. Again we shall find small distinct units, but in this case they will lack firm walls. Nevertheless we call them cells because these microscopically small units which make up the animal body are obviously of the same kind as the cell units in the body of a plant. Whether we look for them in the stem or the root or leaves of plants, in the heart or kidneys, in the blood or bone of animals, Man, newt, or earthworm, whenever we study animals or plants under the microscope we see they are composed of cells — or of a single cell. As we shall see, however, this is not true of the viruses.

Are all these cells living? Is it possible for parts of a living organism to "live" independently, or is life restricted to the living organism as a whole?

Can we split life?

Here some of you will remember that a shoot cut from a geranium, if given proper care, will develop into a new plant, while everyone knows that a limb lost by a bird or frog is dead for

good. But do not let us draw hasty conclusions. Could it be that
in the second case we did not apply sufficient care? Let us ask
Nature herself whether a part of an animal's body can live by itself.

Nature answers our questions if we put them properly, in the
form of a scientifically controlled experiment. Let us take a frog
therefore and treat it with the same consideration a patient gets
from his doctor. It is put under anesthesia. Then it is cut open
and we can see its heart beating. We know it is alive. Now we
take out the heart very carefully. If we just put it on the table it
will dry up and die rapidly. Putting it into plain water damages
it too. If, however, we add to the water a small, carefully calculated amount of common salt and traces of other salts that make
the water as similar as possible to the blood and the body fluids
of the frog, we find that the isolated heart goes on beating for
hours and, if we take special care, for days. This goes to show
that an isolated part of a living body is viable, if one offers it conditions similar to those prevailing within the body from which it
has been taken.

As we know already, the heart consists of many small cells.
Can we carry the splitting-up process further, as it were, to the
cell level without destroying the life of the cells? Indeed it can be
done. Incubated eggs have been opened, and the heart of the
developing young chick taken out and cut to pieces. These, consisting of many cells, were transferred into a carefully prepared
culture fluid. It was found that although the cells went on living
they did not join up again to form a beating heart. Rather, the
opposite happened: they stayed separate. In this fashion they can
live for years, in fact they thrive better singly than they would
in their appointed place, as parts of the beating heart of a bird.

Cells can thus be isolated from a number of different parts of
the body and then grown outside the body. Since we call a community of cells, similar in structure and function, a tissue, we refer
to the growing of cells in this manner as tissue culture. To this
young branch of life science we owe wonderful discoveries,
among them the fact that the cells which make up the living body

are indeed living units. Nowadays it is easy to cut up single cells into small fragments, but even with the most modern techniques nobody has yet succeeded in keeping these cell fragments alive.

3. *Immortal Dwarfs*

Independent cells

Since we are able to keep isolated body cells alive artificially, it is no surprise to find in Nature plenty of cells that live an independent life. They are not, however, cells that have broken free from a body in a bid for independence. On the contrary: the simplest and most primitive organisms among animals as well as plants are single-celled, or unicellular. If it were possible for us to survey the history of our earth over the past millions of years and to watch the genesis of the animal and plant kingdoms, we should discover the unicellulars in the earliest periods of the history of life on earth and we should see multicellular plants and animals evolve from them. But let us look at the unicellular organisms alive today.

Unicellular organisms are minute, for cells are by their very nature tiny, and almost all of them live in water. On land they would dry up very quickly. There exist a great variety of different types, each in enormous numbers. They populate the seas, rivers, lakes, ponds, and puddles, and yet we very rarely notice them, except the plantlike ones which are brightly colored. They can turn a village pond into a grass-green soup, while millions of others can give to water puddles a blood-red color, a sight that has put fear into many a timid and superstitious mind.

Unicellular animals are usually colorless and quite inconspicuous. One of the simplest among them, the amoeba, we find in the mud of ponds. To the unaided eye it looks like a grayish white dot, and only under the microscope does it reveal its continuously changing shape. It moves about, but it may take one to two hours to move a third of an inch. It moves by means of appendages that the zoologists call pseudopodia. The appendages are not permanent structures but arise and disappear continuously. Of course an amoeba does not set one foot before the other as we do, but the feet flow on a set course and the rest of the animal follows (Fig. 2). In this manner the tiny speck of living matter crawls along.

It has no mouth, but by flowing around its prey it eats organisms still smaller than itself, swallowing them up. Any part of its surface that happens to come into contact with food can turn into a stomach. A voracious amoeba may have as many as half a dozen such tiny temporary stomachs going. But these come and go as they are needed and they are generally not to be found in an amoeba which is not engulfing food.

One special structure is, however, always present, not only in the amoeba but in every animal and plant cell: it is a thin-walled spherical or oval body in which under high magnification one can see a few inconspicuous granules; nobody could possibly guess from their appearance what an important role they play in the life of the cell (Fig. 2). The little sphere or oval, embedded in every cell, like the stone in the flesh of the cherry, is called the nucleus.

Life without old age

When there is plenty of food available an amoeba can grow to twice its original size within a few hours. It then gets ready to reproduce itself. This seems to be a fairly simple procedure. The animal stops moving about. Then the nucleus starts to elongate and divide into two equal parts, thus forming two nuclei. The cell itself then divides, and in this manner the mother animal produces as it were two daughter animals, each half the original size (Fig. 2).

In comparison with reproduction in higher animals this is a simple process. On the other hand cell division as such is full

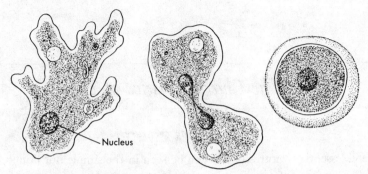

Fig. 2. Amoeba, in movement; in division. Right, the amoeba encysted. Notice the nucleus in division as the single-celled animal divides (middle figure). (Greatly enlarged)

of the riddles of life. We shall not touch upon these now, but rather watch how each daughter amoeba in turn grows up and divides again. Every time, the mother is entirely used up to form her two children; she never grows old nor does she die a natural death, but turns forever into something young again. The same holds true for all other unicellular organisms.

Is it then not true after all that all living beings have to die? Of course amoebae are eaten by their enemies, or the puddles of water in which they live may dry up, or all kinds of other misfortunes may endanger their lives. However, these are external mishaps. It would seem that the amoeba has eternal youth and life infinite.

In these immortal dwarfs we seem to have met something that contradicts the universal mortality of higher plants and animals. But this contradiction can be resolved. Before we do this, however, let us stay a little longer with the amoeba.

4. *The Characteristics of Life*

Artificial living creatures?

The essential features of life can be seen in the simple but doubtless alive amoeba much more clearly than in any of the more highly organized plants and animals. Movement is its first characteristic. But movement cannot reliably be considered a sign of life. Many animals are motionless: the sponge, for instance. Even an amoeba can round itself off and stop all movement for weeks on end. Some decades ago the well-known zoologist Bütschli aroused the excited attention of his colleagues, because he succeeded in making "artificial amoebae": little lumps made of a mixture of oil and soap, which started to crawl and to form pseudopodia like amoebae. With his moving physical models Bütschli could imitate an activity of living material and explain it on the basis of well-known physical laws, but he could not create a living organism. The crawling little lumps of soap and oil can neither eat, digest, nor assimilate, nor can they grow.

Growth and chemical change (metabolism)

Growth as such is not a sign of life either. As children we enjoyed seeing a salt solution evaporate and watching the formation and steady growth of salt crystals. This kind of growth is something very different from the growth of an amoeba. The salt is dissolved and hidden in the solution. What we enjoyed watching was its reappearance in the form of a solid growing crystal. The crystal grows from salt in a regular fashion by mere addition of like to like. When an amoeba feeds it takes up substances that are different from those making up its own body. It digests these substances by breaking them down into components and at the same time making them soluble. Out of these soluble substances it builds up the living matter of its own cell and grows in the process.

Hand in hand with this body-building process go others that are just as important. The amoeba crawls about in search of food. To move about is to do work. We all know that work cannot be done without a source of energy; this is as true for the tiny amoeba as for the biggest steam engine. A motorcar needs fuel if we want it to move, and similarly the amoeba needs a supply of energy for crawling around. It obtains this energy from the chemical compounds that make up the substance of its body and which are continuously replenished by the intake of food. The breaking down of these substances into simpler compounds is a process quite similar to the burning of fuel in locomotives and cars, and it supplies the amoeba with all the energy needed for living. Simple chemical substances that are of no further use to the cell body are discarded. The continuous building up and breaking down of substances by the living cell is called metabolism. If we fully understood this process, the mysteries of life would be solved. We are still very far from this.

The most important compounds of the living cell are the proteins. Proteins have a very complicated chemical structure, and the chemists have not yet succeeded in discovering the finest details of protein structure, though some of the simpler proteins have been synthesized. The structure of many of the proteins which characterize living things, however, still escapes us altogether, understandably so, since a living thing dies as soon as one experiments with it in test tube and flask, and there is indeed little hope that we shall soon fully understand the working of the living cell.

When the amoeba rounds itself off and enters into a prolonged resting stage it stops feeding and moving about and all its manifestations as a living organism are reduced to an absolute minimum. Many unicellular organisms and others too can stay dormant for long periods, sometimes for years, and thus survive unfavorable conditions such as the drying up of their aquatic habitat. In fact they can live without food for so long that one wonders whether their metabolism is not temporarily arrested altogether. It may be that this is so. Are they, therefore, dead? Certainly not. An amoeba, like every other creature, is dead only when the breaking down of substances is no longer compensated for by the building up of new ones.

5. *Let Us Become Familiar with Some More Technical Terms*

We have already mentioned the importance of the distinctive vesicle within the cells of plants and animals that is called the nucleus. The living stuff of which the entire cell is made and in which the nucleus lies embedded is called protoplasm.

When an amoeba produces pseudopodia and crawls about, when it takes in food, digests it, and builds up its body, when it breaks down these substances again in order to get energy like an engine from its fuel, the scene of action is the protoplasm. To what degree the nucleus is involved in these activities is difficult to ascertain, although it may well play a leading role as it does for instance during cell division. Usually it appears to keep in the background. On the face of it the protoplasm is the more obvious seat of the vital activity of the cell. All the more reason for trying to find out about the composition and structure of this wonderful substance.

We know it is composed mainly of proteins, and in the little amoeba we saw the protoplasm "flow." It must be a very special fluid indeed, in which so many things can happen simultaneously, without getting in each other's way, though space is so small. If we want to prepare different dishes, several pots have to be put on the stove. Consider that within a tiny cell such as the amoeba, in comparison with which our cooking pot would be the size of a lake, the production of digestive juice for the little stomachs, the breaking down of food, the building up of body substances, the combustion of the fuel, and many other processes have to happen side by side, all at the same time. What then is the special nature of this extraordinary fluid?

A can of beer helps to enlighten us

When thinking alone does not give us a clue, we start experimenting. Let us take a can of beer from the refrigerator. As we pour it, foam forms on top of the glass. This, of course, is beer too, but with numerous bubbles of carbon dioxide in it. Carbon dioxide is a gas and offers still less resistance than a fluid. But if we stick a match into the foam it will stand upright in it. When the beer goes flat because the bubbles burst, the match topples over at once. Therefore, neither the fluid nor the gas alone could have held it upright. The firmness derives from the peculiar mixture of gas and fluid. We continue our experiment. The firmness of foam shows similarly when we put into it a few drops of ketchup. The red drops stand out sharply against the white foam. They do not merge as they would if we put them into a glass of water. They persist until the bursting bubbles, or our thirst, brings this little experiment to an end.

It is not essential that we mix a fluid with a gas. One can get foams with similar properties from two nonmixing fluids, such as oil and water. Living protoplasm is a foam structure of this kind. This is one of the factors, though not the only one, that allow different reactions to go on side by side, within the limited space of a cell. However, this does not seem to be the only factor, as not all types of protoplasm have a foam structure as we used to believe. The fluid protoplasm has yet other very specific properties, the detailed study of which would involve us deeply in physical chemistry. Let us try to say in a few words what it is all about.

Sugar, medicinal charcoal, and protein

We break a lump of sugar in half, then break the pieces again and again, first with our fingers, next with a pocket knife, until finally, when even the finest instruments cannot divide them any further, we shall have to carry on in thought only. Eventually we reach the limit and arrive at the smallest particles of sugar possible, namely the sugar molecules.

Of course the chemist has the means to break up the sugar still further, but it then ceases to be sugar. He gets down to the chemical elements, the carbon, hydrogen, and oxygen atoms, which compose the sugar molecule. Each of these has quite different properties from those of sugar. Nobody would enjoy a bit of carbon in his coffee instead of sugar, though carbon is one of the elements contained in the sugar molecule.

The molecules are therefore the smallest particles into which a substance can be split without changing the nature of the substance itself. How can we get down to molecules? The experiment is really very simple. All we need to do is to put a morsel of sugar into a glass of water and dissolve it. The water now contains the sugar molecules separately. They are much too small to be seen even under the best microscope, and their number is infinitely greater than that of all the human beings on earth. We cannot catch them with the densest filter paper, because like water they seep through the finest pores. This is what the chemists call a true solution.

For some intestinal troubles the doctor prescribes medicinal charcoal, which comes in tablets and is most easily taken when powdered and put in water. The small carbon particles are quite visible and are, of course, not molecular. When we pour the black fluid through a filter, the carbon stays on the filter paper and the clear water runs through. Here we have a suspension, not a solution. The brown color of our rivers in flood is brought about by a fine suspension of clay and soil particles in the water. One may not notice them with the unaided eye, but they are at once visible under the microscope. Many a town draws its water supply from just such rivers. Large filter systems hold back the dirt and we can enjoy clear drinking water.

But what has protoplasm got to do with sugar dissolved in water and muddy rivers? It is not like either of the two but may be described as something in between. When the chemist splits a sugar molecule he gets a few dozen atoms of carbon, hydrogen, and oxygen, while a protein molecule yields thousands of atoms. The protein molecule is altogether a very much larger molecule though it is still invisible to us and much smaller than the smallest dust particle. It passes through coarse filters but not through the finest. It is this size of molecule dispersed in a liquid that yields what the chemist calls a colloidal solution, or colloid. There are other substances, apart from protein, that can form colloids. The criterion for a colloid is that the particles are larger than in a true solution and smaller than in a suspension. The largish molecules lie so close together that in a colloid an added colored fluid cannot spread out so quickly as it does in a sugar solution with its smaller, more mobile molecules. It is the large size of the protein molecules as well as their foam structure that makes it possible for different chemical processes to go on side by side in living protoplasm without interfering with each other within the limited space of a cell.

Lastly, the molecules have another property important to us. Because they are large, many-branched, and complex in structure, they become easily interlocked or anchored to one another. When this happens the fluid can turn solid. This can happen to fluids other than colloids, of course. Everybody knows that water too can turn solid, namely when it freezes. The specific and very important property of colloidal fluids lies in the fact that the reversible process from fluid to solid and back to the fluid state can be brought about by minute external changes. A small change in temperature, in the salt concentration, or in acidity can turn a protein solution into a solid, whereas such change in reverse turns the solid back to the fluid state. All this explains why the proto-

plasm of a living cell can take on shapes and forms quite unchar-
acteristic of an ordinary fluid.

6. *Multiplicity of Form in Nature's Small-Size Department*

The physicists tell us: a droplet of fluid floating in another fluid
forms a sphere. We leave it to the physicist to work out the
reasons for this. But being conscientious investigators, let us check
this statement by experiment. All we need is a glass of water,
some oil, and alcohol. First, we put a drop of oil on the water.
Oil being lighter than water, it remains on the surface and spreads
out. Being heavier than alcohol, a drop of oil put in alcohol sinks
to the bottom. Next we try to find the right proportions in which
to mix water and alcohol so that the mixture has the density of oil.
Now we shall see that a droplet of oil is suspended in the mixture
and is indeed spherical.

Animals with unlikely shapes

An amoeba resting in water is also spherical. But when we
explore the microcosm of the water we see under the microscope many a shape that contradicts the findings of our experiment.

Perhaps we are lucky and encounter a heliozoan. This is
one of the largest and most beautiful among our unicellular fresh-
water animals and is visible to the unaided eye as a white dot. It
floats motionless in the water. Under the microscope it does not
look at all like our oil droplet but is like a little sun. From all over
its surface radiate fine rays of protoplasm; in short, it presents a
shape that goes against all rules of behavior for a droplet floating

in water (Fig. 3). The explanation is that in the core of the rays the protoplasm has solidified and acts as a support for the fluid protoplasm covering the rays. The protoplasmic supports are not rigid like knitting needles but elastic. They are not permanent like the bone structure in our body but appear and dissappear

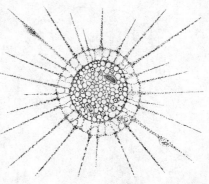

Fig. 3. A heliozoan. (Greatly enlarged)

when and as they are needed. The gentlest touch suffices to make the little sun draw in its rays. We are not always aware of the immediate causes for the change of consistency in the cell, but the observation in itself is a familiar one: we know protoplasm to be a colloidal fluid in which the slightest environmental changes can turn fluid to solid and the other way round. But this is as far as our knowledge goes; the vital process by which the heliozoan grows its rays remains its own secret. We know only that its bizarre shape is maintained by a gel-like protoplasm.

The corona of rays around the heliozoan is in no way a halo. They are poisonous filaments and woe to the little creature that bumps unsuspectingly into them. It will be paralyzed and the fluid protoplasmic cover will convey it along the rays into the body and into its little stomachs, which are known as food vacuoles. This strange creature is a very phlegmatic highwayman, but none the less a very successful one.

Within the same drop of water we may find a much smaller creature not visible to the naked eye. It seems to be in a hurry and moves as if it were pulled through the water by an invisible force. It is egg-shaped and keeps this shape because its outermost cover is a fine membrane of gel-like protoplasm (Fig 4, upper left). What is the driving force of this minute swimmer? The highest magnification shows us at the front end of the animal a

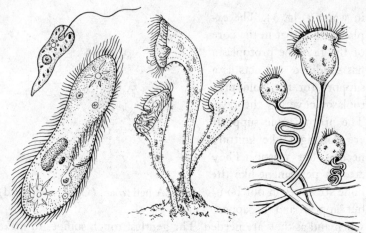

Fig. 4. Variation among single-celled animals (the protozoa). Upper left, a flagellate (Euglena); lower left, a ciliate (Paramecium); middle, the trumpet-shaped ciliate Stentor; right, the ciliate Vorticella. (Greatly enlarged)

thread of exceptional thinness, difficult to see because it moves continuously like the lash of a whip. This lashing forces the water backward and propels the animal forward, just as oars do to a boat. The whip we call a flagellum, and the group of single-celled animals or protozoa possessing them are called flagellates. The hairlike flagella too have inside them a central elastic axis of tougher protoplasm, otherwise they would roll up into a sphere.

A slightly tougher fellow, the ciliate, possesses hundreds of such whips, hairlike structures that cover the whole body (Fig. 4, lower left). They are called cilia. The organism looks somewhat like a little slipper and is about 1/100 inch long. It, too, can keep its shape because of its solidified protoplasmic surface. However, this firm cover prevents it from absorbing food all over its body surface. Only at the very bottom of the groove, which gives the animal its slipperlike appearance, is the surface soft and permeable and used as a mouth opening of the cell. Another special "soft spot" on the body surface is used for the voiding of wastes.

Variety in the internal and external design of
unicellular organisms

By a simple trick one can stop the restless ciliate from racing about, and this allows one to observe some additional wonders. In its interior, at either end, lie two clear vacuoles which alternately contract and fill up again with fluid. The fluid is being collected by star-shaped channels that reach far into the protoplasm. The vacuoles are not stomachs; they remind one rather of a slowly beating heart with which they have, however, nothing in common except their rhythmical activity. They are pumps that continuously remove from the body water that enters partly through the mouth during food intake, partly through the whole surface of the body. But for this mechanism, which is common in other unicellular organisms too, the little slipper would burst.

A sample of water from a well-stocked pool, studied under the microscope, provides an endlessly exciting experience. There is the trumpet-shaped Stentor, which attaches itself by its hind end to water plants (Fig. 4, middle). Inside the bell of the trumpet lies a voracious mouth with a strong crown of cilia which keeps it continuously supplied with food.

One of the most charming and amusing among the unicellular creatures is Vorticella (Fig. 4, right). On a thin, long stalk attached to the stem or leaf of a water plant sits a bell-shaped body and, as in Stentor, its cilia whirl in food-containing currents of water. One could watch it for hours. How useful it is to be able to extend one's greedy body on a long stalk far out into the nourishing food broth — providing there isn't another greedy fellow around. However, in case of danger, the long stalk can contract spirally and bring the little bell close to the protecting platform. The danger past, the stalk expands again slowly and the cilia renew their activity. Such sudden movements are rare in unicellular animals. Under high-power magnification we discover in the stalk a remarkable cord of protoplasm that fans out into the bell-shaped body and which can contract like a muscle fiber. When it does this, the bell is pulled back to the bottom. When the cord relaxes, the elastic covering stretches the stalk and elevates the bell again.

In contrast to the continuously improvising protoplasm of the flowing amoeba, Vorticella is an organism in which different zones of the cell take on special jobs. Thus, the body wall is solid and shape-preserving: the cilia bring in food, the mouth opening is strictly localized, a contractile stalk affords protection; in short, there exists within the single cell a division of labor.

However, the single cell is a very small stage unsuitable for intricate performances on a larger scale. It is in the multicellular plants and animals that Nature perfects the division of labor. There the cells that are produced by growth and repeated cell division stay together to form a cohesive community. In this way are built up the large bodies of a cabbage or an apple tree, of an earthworm or a snail, a cat or a human being. Now whole groups of cells can practice among themselves a division of labor, each group taking on one or the other of the manifold activities of living protoplasm.

7. Ways Toward Specialization and Division of Labor

Scientists believe that some of the very first living organisms on this earth may have resembled amoebae. It is left to speculation to ponder the events that link the advent of those protozoa in the primeval history of life on earth with Man's appearance on our planet. The thought of such a connection is, however, not as fantastic as it might appear. Consider that not so very long ago, in fact just before each one of us was born, our body was one such single cell. Each human being, like every hen, frog, or snail, develops from one single cell, the egg. In the egg state all living beings are still rather similar.

I hear you object: Surely a hen's egg looks different from that of a blackbird, and nobody will mistake the egg of a snail for that of a bird. This is correct but these obvious differences are actually not essential. The shells of eggs do differ and so do their colors. But the shell is only a protective cover produced in the body of the mother. Eggs can be of different sizes, and this depends entirely on the amount of nutriment they contain. The egg of a bird reaches a size that is exceptionally large for a single cell to attain. The reason for this is that within the body of the mother much nourishing yolk, the yellow of the egg, has been deposited within the egg cell, to which a further coating of a large amount of albumen, the egg white, has been added. If we could remove these additional substances we would find that this egg too is but a small quantity of protoplasm, containing a nucleus and looking almost like an amoeba.

When an egg develops into the body of animal or Man, two things happen: first, the egg divides into two cells, both of these divide again to form four, eight, sixteen cells, and so on. Since they do not turn into separate organisms, as in the protozoa, but stay together, multicellular organisms are formed. The number of cells formed varies according to the kind of organism. A small worm is made up of a few hundred, a human being of uncounted millions of cells. The body of an elephant contains more cells than that of a mouse, and it is the difference in the sheer number of cells, and by no means a difference in their size, that turns the elephant into a giant. The second important event happens at a certain stage of development when the cells take on a special shape, related to their future function. They turn specialists. Once more the microscope will help us to understand this relationship between form and function of cells. Let us examine preparations of thinly sliced sections from various parts of the body. They are preserved between two glass slides and labeled according to type

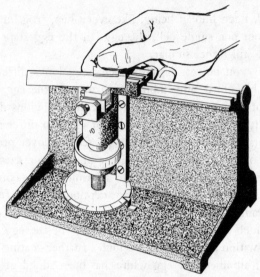

Fig. 5. One type of microtome. A bit of tissue is being sliced with the sharp knife.

or origin. One such preparation is labeled "section through the human skin." It looks simple and yet it took a laborious procedure to prepare it.

How to prepare thinly sliced cold cuts for investigation

If we want to cut thin slices of bacon evenly we use the bacon slicer. If we want to cut pieces of skin or liver thin enough for the cells to be easily visible under the microscope we have to use a microtome. This is a very special slicer with a sharply ground knife that can glide back and forth on a metal slide (Fig. 5). The piece from which the sections are to be cut is raised by a precision screw step by step toward the knife. In this way one obtains sections that are 1/25,000 of an inch thin. For studies under the electron microscope a technique has been perfected by which one succeeds in cutting sections 1/2,500,000 of an inch thin. But let us proceed with the description of the more common routine.

If we try to cut newly baked bread very thinly the slices

easily tear and crumble. The same would happen to a piece of skin on the microtome if it were not properly prepared beforehand. Since it should be as easy to slice as butter, we soak it in a substance that cuts easily. As butter itself would be too soft, one commonly uses paraffin wax, which at room temperature is as firm as frozen butter. In order to make it possible for the wax to penetrate, all the water contained in the piece of skin has to be removed carefully by means of suitable fluids. Next, the skin is placed in a paraffin bath and put into an incubator for the paraffin to soak in. After subsequent cooling, the paraffin block can easily be cut into slices of desired thickness. The sections are then stuck onto glass slides and the now superfluous wax is drained away.

It took much patience and ingenuity to find for each animal and each organ those fluids that preserve the fresh material as true to life as possible, without shrinkage and distortion of the cells. Now at last we may examine microscopically a section through the human skin.

Skin and glands

Whether we take the skin from back or belly, from the hand or the foot, we always find several layers of cells that form the protective covering of the body. They all do the same job and they all look alike. We call such an association of cells a tissue. The task of the covering cells is simple, and they remind us of numerous flattened bodies of amoeba tightly packed together.

In the deepest skin layer the skin cells continuously renew themselves throughout life while the older cells are pushed outward, forming the upper layers. During this progress they die and their protoplasm turns into a horny substance that sloughs off unnoticeably in tiny flakes (Fig. 6). These tough and insensitive cells are extremely well suited to form a covering, and where the horny layer is toughest we can, without incurring any damage or feeling pain, cut a sliver off a fingernail or a horny callus acquired at the oar. When a lizard molts, its covering tissue is shed in large strips, while a snake, in a still more ingenious fashion, just slips out of its skin, shedding the discarded sheath as a whole.

Cells of another type are the gland cells. They produce and secrete specific substances such as mucus, sweat, saliva, gastric juice, bile, etc. The mucus that makes a fish slippery is secreted from gland cells of the skin which are embedded between the covering cells. These are unicellular glands. When many gland cells group together to form a tissue the gland becomes visible to the unaided eye and can sometimes be quite large, like the kidney or the liver. Each gland is a small chemical factory. But so far Nature has not betrayed its manufacturing secrets to us. Nobody has yet found out how the salivary cell makes saliva or a liver cell produces

Fig. 6. Human skin. The outer skin forms a relatively impervious layer in which there is a horny substance. Notice the hair, and the tubular, convoluted glands below the outer skin layer. (Greatly enlarged)

bile. We can only see the coarse structure of the protoplasm, but to observe the molecular structures that hold the gland cell's secret is not yet possible.

A visit to the kitchen

When the cook skins a rabbit, the fine sheets of connective tissue that stretch between the pelt and the meat can easily be torn with the fingers. As the name indicates, it connects different parts with each other — in our case, the skin and the muscles — and it fills gaps. Under the microscope it looks very different from the tissues we have discussed so far. The cells are very small and do not hang together (Fig. 7c). They do not serve as the connecting and filling material. This is done by a substance containing elastic fibers, which the cells have deposited around and between themselves. We find similar arrangements in the supporting tissues of the body, the cartilage and the bone (Fig. 7e, g). It is not the cartilage or bone cells themselves that give firmness to the tissue, but the organic material that they deposit around themselves and which in the case of bone is reinforced by an additional deposit of lime.

Consider the tough, lean mountain climber scaling a peak. His body seems nothing but skin and bone, muscles and sinews. We would have to cut him open in order to find evidence of some fat in his body. Imagine, in contrast, the paunchy figure of a perspiring tourist, and it becomes strikingly evident that nothing, by its sheer expansion, can influence our appearance to such an extent as fat tissue. It establishes itself wherever it can find space, around our internal organs and in many other places, quite especially right under the skin. The fat cells are microscopically small containers, each holding a small fat droplet that pushes the protoplasm and the nucleus close to the wall (Fig. 7h). At first sight they look virtually inert and useless. But let us not look down on them. They are an important store from which our body can draw energy when we have to exert ourselves, and also in times of illness or deprivation. Only when fat is stored beyond our needs does it become a hindrance.

But back to our rabbit roasting in the oven! We like it not for the sake of its fat or bones but for its meat. What the housewife calls meat, the biologist calls muscle tissue. The fibers that are quite easily recognizable in cooked meat are the muscle cells. Almost all their protoplasm has changed into the finest fibrils, which can contract very rapidly and effectively, like the muscle cord in the stalk of Vorticella. The whole muscle cell is involved in the process of contraction, and the longer the cell the more effective it is. In fact it sometimes can reach the length of several inches, which is a size very rare among cells generally.

Many muscle cells are bundled together by connective tissue to form a muscle. Since they are so arranged that they all contract

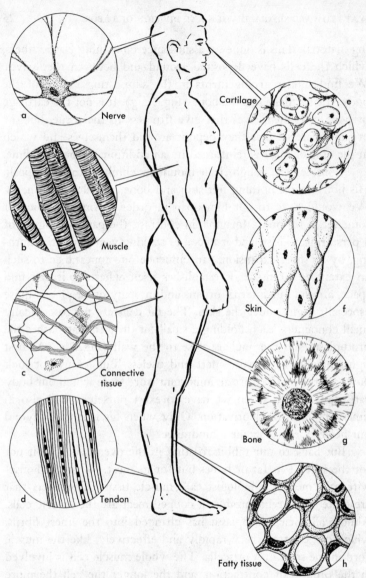

Fig. 7. The human body and types of tissue cells (cells greatly enlarged). Upper left, nerve cell and its fibers; below, in order, muscle cells of the arm, connective tissue surrounding muscle, tough fibrous connective tissue of a tendon. Upper right, cells of cartilage; below, in order, outer cells of skin, structure of bone, cells of fatty tissue.

Nerve a

Cartilage e

Muscle b

Skin f

Connective tissue c

Bone g

Tendon d

Fatty tissue h

in the same direction, their forces add up. When they contract they may for instance move a pair of jointed limb bones closer to one another (Fig. 7b) with the sinews or tendons reaching from bone to bone across the joint. Muscles are attached to bones either directly or by means of sinews or tendons. A tendon is just a bundle of connective tissue fibers that are regularly arranged in a lengthwise direction. Therefore, it is impossible to tear a tendon when pulling it in the direction it normally works. On the other hand tendon fibers can easily be teased apart by a transverse pull, as can be done to the fibers of a bit of string. Once again we find extreme specialization in response to the task of the tissue.

Aristocrats among cells

Of quite a unique shape are the nerve cells, the components of the nervous tissue (Fig. 7a). The brain and the spinal cord, the rulers of the body and the seat of our mental activities, are essentially composed of an enormous number of these cells. In the brain and spinal cord they form a seemingly untidy network of cell bodies, and their fiberlike processes are so intertwined that one would scarcely believe that each fiber has a definite pathway and destination!

Each nerve cell consists of a cell body and a few branched appendages called dendrions, one of which is especially long. The long, fiberlike appendages called axons can leave the brain or spinal cord and join up with their equals to form strands which we call nerves. They can, for instance, run toward a muscle and end at a muscle cell. When we move the muscle voluntarily the command for the contraction of the muscle cells comes from the brain and reaches them through nerve fibers.

In the amoeba the stimulus to move is transmitted by the protoplasm. If one touches a pseudopodium roughly, the neighboring ones are also pulled in. At the point of contact a change in the nature of the protoplasm has occurred, which spreads through the whole cell body. One calls this the pathway of the stimulus. In a nerve cell the stimulus travels very fast and long distances. With the help of their shorter dendrions nerve cells make contact with one another. Here the distances are not so great. The axon, however, which leaves a nerve cell of the spinal cord and transmits the command to contract to a muscle in our foot, in the human body is about one yard long and in the elephant is even longer.

These are extraordinary dimensions when we compare them with the usual size of cells. In earlier days it was doubted whether this long axon is really a part of the nerve cell or rather the product of a number of cells making contact. The American scientist Ross Granville Harrison, who was the first to grow tissue cultures, was successful in observing, under the microscope, the growth of these processes from young isolated nerve cells of frogs and could show that they are undoubtedly part of the nerve cells.

Sensory cells have structures related to their functions. The protoplasm of an amoeba can be stimulated by various external factors such as touch, strong light, chemical substances, and so on. In the sensory cells this original general "excitability" is superseded by an enormously intensified power of reaction to specific stimuli. Thus the sensory cells in our eyes respond to light stimuli, the auditory cells in our ears to sound waves, the olfactory cells in our nose to certain chemical substances. As a specialist in its own field of activity, each of these cells works more efficiently than the cell of an amoeba ever could. Specialized as they are, they would, if left to isolated existence, be useless. In their proper place in the body they are intermediaries between the external world and our internal experiences. Specialization is not a shortcoming but an asset, provided it is used to work toward greater perfection of a whole.

8. *About Body Cells and Germ Cells, Death and Immortality*

To Man, life without death would seem to be something so perfect that faith and myth consider it to be the prerogative of God
and godlike beings only. Strangely enough, however, it is in such
simple and primitive organisms as the protozoa that potential
immortality is glimpsed. They are not immortal, it is true; many
of them die in the struggle for existence, against their own kind
and against other forces. But they carry within themselves the
possibility for unlimited life and if they do not suffer an individual
mishap they do not die of old age. Yet in higher plants and animals, including Man, who considers himself to be the crown of
creation, life irrevocably ends in natural death. It looks as if this
might be the fundamental difference between the simpler and the
more developed organisms, seemingly leaving the latter at a disadvantage. Things are, however, not what they appear to be if we
look at them the right way.

At the root of natural death

Among the protozoa we got to know the dainty Vorticella (Fig.
4). In spite of the enterprising shape of its body it multiplies in
just as simple a manner as the amoeba. First the nucleus divides,
then the bell, and finally the stalk splits lengthwise. Thus one can
find on the leaf of a water plant hundreds of animals sitting together, all of which have grown from one another by such divisions. Now each one is quite independent and leads its own life.
In a related species of Vorticella the stalk does not divide right
through, with the result that after repeated divisions many heads

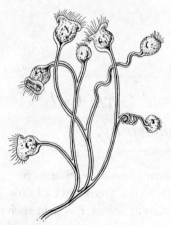

Fig. 8. A colony of vorticel-las (greatly enlarged). The common branched stemlike structure joins the proto-zoa.

sit on a common branched stalk (Fig. 8). They are all connected but are nevertheless not a multi-cellular animal but a colony of uni-cellular ones. All are alike and each little bell, if conditions turn un-favorable, can leave its stalk, swim away, and settle down elsewhere to form a colony of its own. Again each of these organisms can live on indefinitely, like an amoeba.

Among flagellates, too, we find forms that stay together after divi-sion to form a community of cells. Here also the cells are in most cases alike and capable of reproducing in-definitely. The largest and most beautiful among them is called Volvox. Its body consists of about 10,000 flagellates and keeps a remarkable balance between being a colony of unicellular animals and a multicellular organism (Fig. 9). Its cells are not alike any longer; those at the forward-facing pole of the spherical body have stronger flagella and are more sensitive to light. We encounter here a first indication of a division of labor which expresses itself even more strikingly in the fact that not all but only a few cells are capable of division. When they do so they form daughter spheres that grow inside. The mother sphere busts and, liberated, the daughter spheres start their own life. This is the first cell community where we find, side by side, cells of limited and of unlimited duration of life — the simplest organism that dies a natural death.

The separation into body cells and germ cells

In multicellular animals the process is essentially similar. A far-reaching division of labor has produced the development of the different cell types described in the previous section. They are specialized to perform a limited task to perfection and in doing so they wear out and eventually die. In some cells the signs of increasing wear are clearly visible. For example, in the nerve cells of aging animals or human beings, dark granules and bodies are being deposited in increasing numbers while the active living protoplasm is shrinking. No wonder the cells eventually cannot function any longer. And when, in a cell community (an organism) where everything depends on a balanced interaction of the parts, only one single important group of cells stops functioning, the fate of the organism is sealed.

It is not always the nerve cells that grow old first and thus end life. Sometimes other important cells take the lead along the path of decline. We do not know much about this. We cannot even assert that they must inevitably grow old and die, unconditionally and under all circumstances. Provided we remove them early enough from the community of cells and grow them under most favorable conditions in tissue culture, they live on and multiply indefinitely. Under natural conditions, however, left in the service of the body, their fate is wear and tear to the bitter end. All those cells of specialized function and limited life are called body cells or somatic cells. Their breakdown brings about natural death.

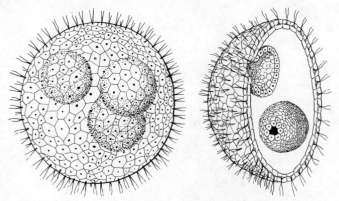

Fig. 9. Volvox with daughter colonies. (Greatly enlarged)

One group of cells, however, is an exception. They are the germ cells. They do not turn into specialists and take no part in the strenuous daily chores of the body. They retain all the manifold faculties that are hidden in the egg cell and become the origin of the next generation. The female germ cells are eggs, and those of the male spermatozoa. This, however, does not interest us much in this context. Indeed, there are animal species in which only the female sex exists and others in which males are a great rarity. When a multicellular animal develops from an egg, one group of cells is always set apart. These cells do not turn into skin cells or gland cells, into muscle or nerve, but remain unspecialized and omnipotent. They show no signs of age and, as germ cells, carry the potentiality of life from generation to generation. This is the biological proof positive of the saying that parents live on in their children. It is literally true.

II

THE ORGANS OF THE BODY
AND THEIR FUNCTION

We have learned to know cells as the smallest units composing the bodies of organisms. There exist, however, organisms of a higher order which are each composed of thousands of cells. When we dissect a large animal these more complex units inside it are very conspicuous, and we call them organs. The name comes from the Greek word "organon," which means tool. They are, so to speak, tools used to perform certain tasks in the service of the body as a whole, and they form parts of the body distinctly different from each other and each functions in a special way. Thus the heart is the tool for the transport of the blood, the kidney for the excretion of urine; stomach and intestine are involved in digestion, the muscles bring about movements, the nose is a tool for perceiving smells, and the eyes are wonderful instruments through which the beauty of a landscape finds its way into our mind.

Usually there are several tissues involved in the formation of an organ. The intestine has an inner lining of covering tissue; besides, it contains glandular cells that produce the digestive juice, muscle cells that move along the food contents inside it, nerve cells that regulate its movements, and connective tissue that holds everything together.

Each organ is a well-balanced unit of a higher order, a unit that cannot exist on its own but only in co-ordinated function with the other organs of the body.

In order to understand the functioning of the body as a whole we have to study the organs, just as we have to be familiar with the parts of a machine if we want to understand how it works. Let us begin with the seemingly hardest and driest of all the organs, our bones.

---

1. *About the Bones of Our Body and Other Skeletons*

Bones have to be hard. They support and protect the body. The skull, a hard bony capsule under the scalp of our head, protects our brain. If it were not for it, the fist of a child could kill a giant because the brain could not stand the blow.

Most bones, however, are essentially supporting structures. The backbone keeps the human body from collapsing into a heap. It consists of about two dozen single bones, the vertebrae. These alternate with thick connective tissue disks and thus build up our backbone or spinal column (Fig. 10). It has a certain degree of flexibility, otherwise we should stand as stiff as a poker. The spinal column and with it the body are supported by the legs, fitted into a pelvic girdle of bones. Long bones support the legs, and these are movable at the joints. So are our arms, and we can easily feel one bone in the upper arm, two in the lower arm, the wristbones, the five bones of the middle hand, and those of the fingers. Then there are the ribs, which encase and support our chest and help with the breathing. We thus have been introduced to the essential components of our skeleton but not yet to their properties.

Bones are hard. But is this all there is to them? To answer

this question we can carry out an experiment that is as simple as it is instructive. We put a bone, not necessarily a human one, for some time in a container with diluted hydrochloric acid. We know the acid dissolves the carbonate of lime and we expect the bone to dissolve like a piece of sugar in water. This it does not do! It retains its form unchanged because it does not consist of lime only but also of a mass of connective tissue among which the lime has been deposited. The leftover connective tissue is so soft that we can now bend the bone between our fingers, and at the same time it is so elastic that it returns to its former shape when we stop bending it. We put another bone in the fire, which will burn away the connective tissue. The form of the bone again remains unchanged as the heat does not destroy the lime. But now the bone is so brittle that it crumbles when we break it between two fingers. It is the combination of the two components, the elastic foundation and the brittle lime, that makes the bone such a suitable organ of support. It spells trouble if the two components are not present in the right proportions. If there is a lack of lime the bones bend under the weight of the body and we get rickets, a familiar children's disease. When in old age the connective tissue part shrinks, the bones become brittle and a fall can lead to dangerous fractures.

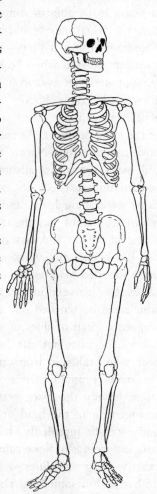

Fig. 10. Human skeleton.

The design of bones — a feat of mechanical engineering

It is proverbially difficult to please all parties. This holds not only in human society but in living nature as well. To find a satisfactory balance between contrasting needs concerns us in this case to the very marrow of our bones. The bones have to be strong in order to be able to support the body. There comes a point, however, when they would become so heavy and massive as to be a hindrance by their very weight. The builder is confronted with a similar problem when he has to construct a balcony that is to be supported on cast-iron columns. Now, the engineers have discovered that a hollow column with the proper thickness of its wall can carry almost the same weight as a solid column of the same diameter. At the same time material is saved and the price is kept down. Nature is the inventor of the hollow column and made use of its advantages uncountable years before the human mind existed: the swaying stalks of grasses and reeds carry astonishing loads, and so do the long bones in our limbs, which are hollow like the columns the builder uses and combine saving in weight and material with undiminished efficiency.

But better still, when a column has to carry not a vertical but a lopsided load, as for example in a crane, its different parts are exposed to very unequal strains and stresses, the direction of which can be worked out by calculations and experiment. The engineers speak of lines of stress and tension within the column besides the parts not affected by the load. Therefore, the engineer who builds an iron crane acts purposefully when he puts the struts economically along the direction of these lines of stress, leaving the space between them empty. We see a similar arrangement in the head of our femur, which has to carry our body weight lopsidedly. It was a surprise to discover that the bars and strands of bone running through its interior are arranged exactly along such lines of stress. A cunning engineer could not do better. But something else has been accomplished which even the most efficient builder can never achieve: when, by some misfortune, a bone breaks and, as sometimes happens, heals together unevenly, the lopsided load will now exert itself in a slightly changed direction. Then the structural bony bars become realigned along the direction of the new forces. Now, during the rebuilding of the bone it is found that thus are combined rigidity and plasticity! We shall find over and over again that the living body as a whole and in its parts works with an efficiency that has the appearance of purposefulness. Many a thinking mind has been struck by this trait, which can be followed through the whole of the animal and plant kingdoms.

A side issue

Is it wrong to speak of plant and animal kingdoms only and not refer to the human race as a third and superior group? Certainly not in this context. The skeleton of an ape is in all its details extraordinarily similar to the skeleton of a human being. The difference between them is much smaller than the difference between the skeleton of a dog and that of an ape, let alone that of a fish. All other organs show this similarity too. Among scientists there is, therefore, no doubt that Man in his organization belongs to the animal kingdom, and nobody need be embarrassed about the features we have in common. It is in the sphere of the mind that we differ most from other animals.

The comparative study of the similarities and differences in the basic plans of bodies is a very attractive occupation, especially since we now know that the animals and plants alive today did not always exist in the form known to us, but evolved quite slowly and gradually toward it. Their build is testimony of their history. We do not want to go into details just now, but a cursory glance at the illustrations of the different skeletons, showing the skeletons of Man, ape, dog, hawk, salamander, and fish (Fig. 11), illustrates, in spite of graded differences, the inherent relationship in the basic plan of design. All have a spinal column. Man and the apes have arms; the dog has none. But its fore limbs correspond to our arms: we find in them the same bones, but they are

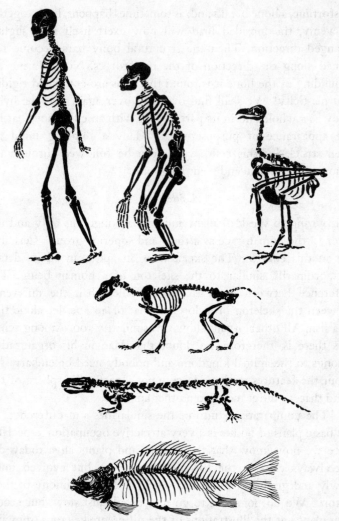

Fig. 11. Comparison of skeletal structures of the vertebrates: Man, ape (both are mammals and primates), hawk (bird), dog (mammal), salamander (amphibian), and fish.

used for standing and walking as in most mammals. The birds, with an upright gait like Man, have wings instead of arms. Here again we find the same bones but slightly altered for different use. Even the fish still shows the fundamental similarities, and its pectoral and pelvic fins correspond to the fore and hind limbs of the salamander and other land animals.

Knights in armor

Has a beetle got a skeleton too? Not everybody will know the answer. The zoologists, however, know it and tell us that a skeleton similar to our own is found only in fish, amphibians, reptiles, birds, and mammals, and since they all have a spinal or vertebral column we call them vertebrates. Insects, crustaceans, snails, and other lower animals have different skeletons. In fact skeletal structures of the most varied kinds are fairly common. Nature is a prodigious inventor.

The beetle does not have an internal skeleton like Man or like a vertebrate. It has an external skeleton in its very skin. That is why it feels hard to the touch. The skeleton does not consist of bone, since this would make the little body too heavy for flight. It is made of a remarkable substance called chitin, combining great firmness with a minimum of weight. The chitin is secreted by the skin tissue, and the body is thus surrounded by a protective covering resembling the armor of a medieval knight (Fig. 12). It is jointed in the same way, to guarantee freedom of movement. All arthropods (crustaceans, insects) have such a jointed armor. This kind of skeleton keeps its shape even after the death of the animal. It is, therefore, easy to keep collections of dead insects. One could not pin up a frog in the same way — it would lose shape and decompose right down to its skeleton.

There is one disadvantage to the otherwise perfect external skeleton: it cannot grow and it cannot be expanded. At certain times, therefore, the armor splits and has to be discarded. This process is called molting, and it is a dangerous event. It is not easy for a shrimp to pull its fine feelers or legs out of their long skeletal tubes undamaged. After a successful molt the new skin is very soft and vulnerable. During this period the animal grows

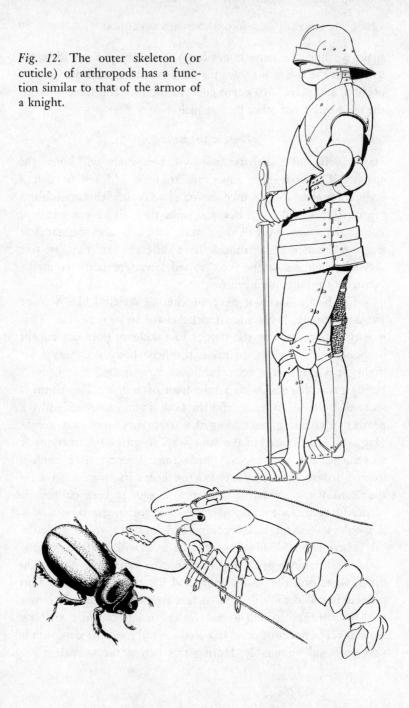

Fig. 12. The outer skeleton (or cuticle) of arthropods has a function similar to that of the armor of a knight.

considerably and quickly, until after a few hours the new covering has hardened, to give protection. This puts an end to growth until the next molt is due.

Many tiny animals build mountains

Skeletons exist in many other shapes, and much could be told about the protecting shells of snails and mussels, of the spiny armor of sea urchins, of the calcareous foundations of corals or the glassy scaffolding of some bizarre deep-sea sponges. The most remarkable fact is that some of the microscopically small unicellular animals build most efficient skeletons. In the sea we find an abundance of amoebalike organisms called foraminifers. They store in their protoplasm carbonate of lime, traces of which are always contained in the sea water, and then deposit it on the surface of their body in the form of a delicate shell, which in some species reminds one of a little snail's house (Fig. 13).

The shell protects the tender body but is in the way during cell division. During reproduction, therefore, the old shell is discarded and later a new one is formed. This explains why in places where the animals are common we find such enormous numbers of their shells. A careful observer will discover that, under a magnifying glass, what he first took for grains of sand on the beach will often be empty shells of foraminifers. Through

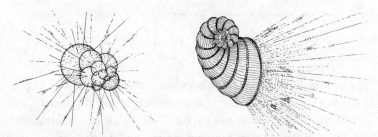

Fig. 13. Foraminiferans. The slender threads of protoplasm (pseudopods) protrude from the protecting dainty calcareous shells. With these pseudopods the animals move and capture food as well. (Greatly enlarged)

the activity of these organisms over countless years, enormous quantities of carbonate of lime have been extracted from the sea and thus deposited.

Since during the history of our earth parts of continents have sunk into the sea and alternately parts of the bottom of the sea have been raised out of the water, we can find such deposits of earlier periods now well inland. Layers upon layers of mighty mountains consist of limestone, which in prehistoric times, when those parts were covered by sea, were deposited by foraminifers. Eventually the uncountable numbers of shells baked together and formed the firm rock in which they are still recognizable today.

The foraminifers are by no means the only agents of rock formation. Coral reef owes its existence to the activity of coral polyps of days long past, while today, in the southern seas, their descendants still go on building ever-growing reefs, a threat to many a ship. Mussels, snails, and other similar animals have contributed their shells to the formation of limestone, and some other members of the plant kingdom have added their special and substantial share. However, foraminifers, in their instinctive activity, are among the smallest and most accomplished builders that have helped to shape the earth's crust.

2. Organs of Motion

Many a medical student has bemoaned the fact that the human skeleton consists of so many bones, which he has to identify and memorize. Could Nature not have made our skeleton in one piece?

Of course anything can be done in several ways. Supporting tissue can be of one piece, such as the wood of the trees. Without wood a beech tree would collapse. We find no subdivisions into single parts of wood in it, but then a tree need not move. Some animals also are immobile. The silica sponges of the deep sea, the tender body of which is supported internally by a glassy scaffolding, or the corals with their calcareous skeletons neither move nor retreat from danger.

Our body is capable of movement because our skeleton is constructed from many bones linked by joints; the organs that set our bones in motion are our muscles. Our muscle cells can contract lengthwise, and this they do very well, very quickly and energetically. Indeed, this is all they can do. They cannot pull the bone to which they are attached first in one and then in the opposite direction. The many movements of which our hand is capable are brought about by the combined action of differently arranged muscles. Our hand alone is served by fifty different muscles. The violinist who plays a sonata masters the activity of those fifty muscles of his hand to a nicety, without knowing them or having to think about them. To study their arrangement and interplay would be quite a long task. We choose a simpler example.

Among the most familiar of muscles is the biceps. It functions when we raise our lower arm at the elbow joint, for instance in the act of lifting. Each muscle needs a counterpart. The counterpart of the biceps faces it at the back of the bone in the upper arm, and when it functions it stretches the lower arm from the elbow joint (Fig. 14). When one muscle contracts its counterpart relaxes, and vice versa. Between them they effect the position and movement of the parts of the limbs that they connect. The joint between the bone supporting the upper arm and the inside bone in the lower arm is a hinge joint. This means we can move the limb in one plane only, like the blade of a pocket knife. One muscle and its counterpart are sufficient to move such a joint. Other joints, like the ball and socket joint in our shoulder, have more freedom of movement, and this joint is worked by nine muscles. Greater freedom, however, means greater risks. It is not just a coincidence that we hear more often of a dislocated shoulder than of a dislocated elbow.

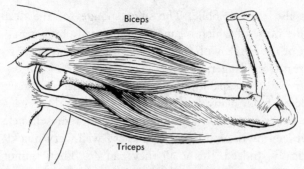

Fig. 14. The biceps, the triceps (its counterpart), and the articulation of the upper and lower bones of the arm.

Walking and motoring

Everybody knows that the joints of a machine have to be well oiled in order to prevent them from heating up and wearing out. Nature has equipped our joints with an excellent lubricating substance which is continuously secreted into the joint capsule in sufficient quantity to cut down friction to a minimum. Therefore, our knee joints do not heat up during a long hike.

Our whole body, however, gets warmer when we move about vigorously for some time. The source of this heat is the activity of the muscles. We know that our car stops if we run out of gas. To move a car means work, and work needs energy. We obtain this energy by burning gas. When a substance burns it combines with oxygen, and heat energy is released. This process is called oxidation. However, part of the heat of oxidation of a fuel, such as gasoline, can be made to move our car, and then we speak of the release of energy.

The development of heat during oxidation can be so violent that small particles tear loose from the burning substance and start glowing in the heat, and thus form a visible flame. The same chemical changes can take place at very slow speed: then we do not get any glow and flames but just a gentle development of heat. Such slow oxidation occurs in the cells of a working muscle. Muscle fuel, the source of muscular energy, is generally glucose, a sugar. This may sound surprising. It would, however, be quite wrong to think that you need only to eat a lot of sugar in order to grow strong. As we shall hear presently, bread and potatoes too are changed into sugar by our digestive organs, and so the blood contains plenty of dissolved sugar with which to supply our muscles. This sugar solution would soon be lost from our body if the muscle cells could not retain it. They do this by quickly converting several glucose molecules into one molecule of glycogen, which is now insoluble and can be stored in the cell. Glycogen extracted from a muscle by chemical means appears as a white powder that looks like flour. This then is the store of energy used by muscle cells in somewhat the same way as gasoline is used by the car engine.

Why we get hot when we are running and why we get out of breath

For a while it was thought that muscular activity was dependent solely on the energy released by the slow combustion of glycogen. This thesis has now been discarded. There exist other chemical reactions that release energy, reactions in which oxygen intake is not involved and which are, therefore, not oxidation processes. The breaking down of certain phosphoric acid compounds plays a most important role and is now regarded as the immediate source of muscle energy. It is now popular to refer to these compounds in the daily newspapers. For instance, you may read of ATP. ATP is an abbreviation of adenosine triphosphate — one of the phosphoric acid compounds to which we referred.

Nevertheless there is an oxidation process going on in our muscles. The active muscle cells would soon exhaust the store of the just mentioned phosphoric acid compounds if their breakdown were not kept in balance by a continuous resynthesis. The considerable energy that is needed for this resynthesis is obtained by the oxidation of glycogen, breaking down to glucose, in the muscle. The glycogen, via intermediate stages, is eventually oxidized to carbon dioxide and water.

The details of the processes involved are fairly complicated and not yet completely elucidated. But we now know enough to understand some common everyday experiences. We know that we have to eat in order to supply our muscles with the source of energy that is derived from food; that we feel warm when we move, because the reactions and oxidations in our muscle cells liberate heat; that we get short of breath when we walk quickly or for a long time because a raised rate of oxidation in our muscles means a raised rate of oxygen consumption and a greater demand on the intake of air into our lungs; that we get tired from bodily exercise and that the muscles eventually refuse to work, because their store of glycogen is slowly used up and a period of rest is necessary for the replenishment of it.

The main question — How does the muscle contract? — we have not yet answered, and for good reasons. There exist several hypotheses attempting an explanation, but no sure proof. We know the fuel used by the muscle cell, we know fairly well how it is transformed into energy. But the exact process of transmission by which energy is turned into contraction of the muscle cells and thus into motion is still the cell's undivulged secret.

Movements in animals and plants

So far we have only talked about our own organs of motion. If we attempted to describe generally the modes of motion in other living organisms it would make up a big chapter. There are the crayfish and insects, with their hard external skeleton which harbors the soft parts of the body including the muscle. Here again a muscle and its counterpart bring about the movements. Here, too, the type of joint determines the freedom of movement of the limbs. Then there is the earthworm, without bones or skeleton of any kind. It moves forward with the help of the many muscles in its body wall, by alternately contracting and stretching parts of its body. It is in no hurry. Its muscles contract relatively slowly, while the muscle of a fly can contract within

1/1,000 of a second. That explains why we can easily catch an earthworm but not a fly.

There exists movement independent of muscle cells. We know already about the movement of the protoplasm in an amoeba and about the cilia of the slipperlike protozoa. Nearer home we find the same special modes of motion. Certain cells in our blood creep about like amoebae, and the inside of our windpipe is covered by cilia which move choking slime upward and out of the lung. Here too glycogen supplies the energy for movement, only the motor looks different.

Now, plants move in a different way altogether. They are not really as inert as they appear at a first glance. A shoot that turns toward the light is said to be growing toward the light. This movement is in fact brought about by growth. Since the unilluminated part of a plant grows more quickly than the illuminated, the plant bends toward the light. This is a perfectly legitimate way of movement for a living organism that is growing and has plenty of time to do so.

Not all plants are so slow. There are the rare instances that perplex us by a very sudden change of appearance. When we touch certain mimosa plants, all the leaflets suddenly close up and the whole leaf droops. Very quickly all the leaves of the plant follow suit.

The mechanism that brings about these changes is again a different one. The green parts of plants exhibit a certain firmness because their cells contain a fluid — the cell sap. Each single cell is hard and stiff with it like a blown-up balloon. The botanists say that such cells have a high turgor. A cut-off branch or a potted plant that we do not water properly withers. This means the water that evaporates from the cells is not replaced; in other words the turgor diminishes, the cells turn soft and flaccid, and the leaves droop. The reactions of a mimosa plant are based on changes in turgor. We do not, however, know how the plant does this. There is still plenty to do for generations of eager biologists.

3. Our Daily Bread

Fuel and building materials

We already know that, like a machine, a living organism needs a supply of substances (from food, air, water) in order to work. Unlike a machine, however, the living body has to build up its own structural parts. Therefore, not only fuel but building material is needed, especially during growth and afterward for general maintenance. Cells are continuously used up, die, and have to be replaced. Experiments with radioactive elements were needed to give us insight into the extent of this turnover. We use artificially produced radioactive carbon, phosphorus, and other substances that are important for the formation of living cells. The elements are not changed in their chemical behavior by this treatment, but they now send out radiation that can be traced by means of sensitive instruments.

Thus a food substance, such as a certain protein, can be "labeled" by building radioactive carbon into it. Traces of this radioactive protein are taken up with the food and eventually built into the cells, where they make their presence known by their radiation as long as they are present. In this way we found that within five months about half of the cellular protein of the human body is broken down and renewed. This shows that considerable quantities of building material are needed to repair tissue and produce new cells. Animals like butterflies and some other insects have a very short adult life and can manage on fuel foods alone. Their source of energy, therefore, needs to be just a simple basic substance. A butterfly can live on the sugar solutions contained in the nectar of flowers. Other animals use fat as fuel. Neither sugar nor fat contains nitrogen. Yet the most important components of protoplasm, the proteins, are nitrogen compounds. They are therefore the most essential building materials.

We have now enumerated the three groups of chemical compounds among which are the most important food substances for human beings and the majority of animals: carbohydrates, fats, and proteins. When we eat an egg sandwich we treat ourselves to all three of them.

The basic foodstuffs

The word "carbohydrate" means that each molecule of such a substance contains carbon, hydrogen, and oxygen, the latter two in the same proportion as in water. The water molecule consists of two atoms hydrogen and one atom oxygen, in chemical shorthand H_2O. Sugar has the formula $C_6H_{12}O_6$, which means its molecule consists of six carbon, twelve hydrogen, and six oxygen atoms. We have learned that in the muscle cells glycogen is formed by the combination of several sugar molecules. This places it among the carbohydrates.

In the life of plants sugar and similar compounds play just as important a role. Plants too can combine sugar molecules to form something like glycogen. They store this type of carbohydrate in their cells in the form of starch grains. These are found, above all, in tubers and seeds from which new plants are going to develop (Fig. 15). Here they are needed most. That is why potatoes and grain have a high starch content. Potatoes and flour are, besides sugar, the mainstay of our larder.

Fats contain the same ele-

Fig. 15. Grains of starch in the cells of the potato. (Greatly enlarged)

ments but in different proportions. The American housewife uses mainly animal fats such as butter, cream, and bacon; the Italians use olive oil, a vegetable fat. Each country makes use of what Nature offers most plentifully.

The proteins have a much more complicated structure than the carbohydrates and fats. Apart from carbon, hydrogen, and oxygen they always contain nitrogen, usually some sulphur, and phosphorus too. One of the simplest of all proteins, clupein, has the formula $C_{179}H_{405}O_{69}N_{99}$. Seven hundred and fifty-two atoms are involved in the formation of one single molecule. The protein molecules of the protoplasm are at least four times, occasionally several hundred times, the size of clupein and are therefore really giant molecules. The presence of nitrogen is mandatory as it is their most essential component. We introduce nitrogen into our body mainly by eating protein (meat or fish). We obtain it therefore from proteins that other animals have built up in their muscles.

The value of food is often expressed in terms of calories. *Calor* is the Latin for heat, and a calorie is the quantity of heat necessary to raise the temperature of one liter of water by one degree centigrade. The physiologists have found that one gram of sugar or protein produces four calories on being oxidized in the body. This amount of heat could raise the temperature of four liters of water by one degree centigrade. One gram of fat produces about nine calories. The heat of combustion set free by any kind of food can be determined, and calorie tables are found in books on dieting.

Of course not all our food is changed into heat. Some of the energy is used for the business of living, in work and play; some of the food is used as building material. The heat of combustion simply indicates what quantity of heat could be set free if the substance were completely oxidized. It is a useful and instructive means of assessing the comparative value of different kinds of food.

A grown-up person, who does not work hard and wants

neither to put on fat nor to lose weight, needs a daily food intake equal to 3,000 calories. Millions of human beings now living know what it means when the daily ration corresponds to 1,300 calories only. They do not need to be scientists to know that this is not enough. In the healthy body the pangs of hunger demand that quantity of food which corresponds to the daily need with admirable certainty and commanding insistence, so that we can expect, without having to weigh ourselves or our food, that our body weight will remain the same for years.

Vitamins

Not so very long ago it was thought that all an animal needs to thrive on are carbohydrates, fats, and proteins, together with water and a small quantity of minerals. This, however, turned out to be wrong. On trying to bring up young rats on substances from the above groups of a high degree of chemical purity it was found that they just would not grow. Only after small quantities of their natural diet had been added to their food did they grow into normal strong youngsters. From these and other experiments it was concluded that, apart from the known components, well-balanced food seems to contain a small quantity of additional substances absolutely essential for life. These were called vitamins, and to date we know about twenty different ones. The chemical composition of some is by now well known and we can even synthesize them. We find them especially in fruit and fresh vegetables, but some of them are destroyed during cooking and canning.

We know very little about the reasons for their beneficial effect. We do know of some that help to build up vital compounds used by the body. In most cases we are much more familiar with the damage that ensues when they are absent in the food. The damaging effects have been known for a very long time, but we could not explain them. The lack of vitamins not only stops normal growth but brings about serious illnesses. Members of polar expeditions, who had to live for months on end on a fairly monotonous diet of canned food, suffered severely from scurvy. In scurvy, bleeding of the gums, ulcers, swellings, even internal hemorrhages cause intense pain and the disease can lead to exhaustion and death.

It was known that even quite small quantities of fresh vegetables, tomato, or lemon juice will stop the disease. It is fairly recent knowledge, however, that these substances contain vitamin C, the absence of which causes scurvy. For convenience the letters of the alphabet are used to name the various vitamins. Shortage of vitamin D causes rickets, a metabolic disease that has to do with insufficient deposition of lime in bones. Cod-liver oil, which is extracted from the livers of codfish, is rich in vitamin D. Vitamin D is stored in the liver fat of fish and uncountable spoonfuls of beneficial oil are swallowed by children, sometimes with considerable disgust. Originally vitamin D is produced by minute siliceous algae. They are eaten by small marine animals and the vitamin finally finds its way, sometimes via several stomachs, into the fish. It is also directly available to us in fresh vegetables and fruit, alas, in an unfinished and therefore useless form. In our body it can become converted into its active form by exposure of the skin to sunlight, especially at high altitudes. This explains why a holiday high up in the mountains can be as good for us as oil from the depth of the sea. The necessary quantities are minute. Rickety rats get well when fed daily on 1/30,000,000 of an ounce of vitamin D. No wonder that it took us a long time to become aware of the existence of these important food substances!

If any one of the vitamins is entirely absent from our food for a certain length of time typical deficiency diseases will befall us. Apart from the specific consequences, a diet poor in vitamins generally leaves the body more susceptible to infectious diseases. It is a sound instinct that makes us like fresh fruit and vegetables, and we normally take up a sufficient supply with our food. Only foolish food fads or stark famine could deprive us nowadays of a naturally balanced diet.

We are vegetarians, after all

As we know, Man lives on a mixed diet. This means we live on vegetable and animal foodstuffs. If we look carefully into this we become aware that ultimately plants are the source of all our food. Cattle are vegetarians, we eat their meat, and thus we live actually on the grass of our pastures.

The same is true of the exclusive meat eaters, the carnivores. The fox that kills a rabbit feeds on the meat, fat, and glycogen that the little rodent built up by eating cabbage leaves and other plants. If our fox eats a mole, which itself is a carnivore living on earthworms, insect larvae, and other such appetizing things, the detour is just slightly longer; with the earthworm and caterpillar we return full circle to the herbivores.

All the food of animals is thus derived directly or indirectly from plants. This has a deep significance. In order to understand it we need to know how plants feed.

How plants feed

Nobody has even seen a green plant eat. Does it suck its food from the soil with its roots or does it live on air? Well, both are true. The moist soil supplies water with mineral salts dissolved in it, and carbon is taken from the air. This sounds like a tall story. Who would believe that the smoky air of our towns and the fresh sea breezes alike contain carbon, the element that we burn as coal?

The air everywhere contains small quantities of a gas, carbon dioxide, a chemical compound of carbon and oxygen (CO_2). Carbon dioxide and water are the raw materials from which sugars are built. And from the sugars starch is produced and stored in the plant. In this process, called photosynthesis, oxygen is liberated. For this chemical synthesis from carbon, hydrogen, and oxygen, energy is needed. This energy is stored in the newly formed chemical compounds, and during their later breakdown it can again be released as heat or body movement. The energy for chemical synthesis in plants is derived from sunlight, and we therefore speak of photosynthesis. Chlorophyll, the green pig-ment of the leaves, absorbs the part of the sunlight that is used as the source of energy. There is very little carbon dioxide in the air and yet it is so important for plant life. Roughly one fifth of the air is oxygen while nitrogen and certain rare gases make up the rest.

Plants can be grown from seeds when their roots, instead of being in the soil, are immersed in water containing the salts nec-essary for their development and growth. If we omit all nitrogen compounds from the water the growth of the plant will stop as soon as the reserves contained in the seeds are used up. The plant will grow and flower only if we add, apart from certain other salts, nitrogen compounds such as potassium nitrate. Thus the ni-trogen from the soil and the carbon from the air are the two most important elements from which the green plant builds up the proteins of its protoplasm.

Animals cannot assimilate or photosynthesize like the green plants, nor can they produce any organic substances from simple raw materials taken from the soil and the air. This is the reason why green plants are the source of all food. All animals are depend-ent on organic compounds which green plants procure with the help of sunlight. There are even plants devoid of chlorophyll, such as mushrooms, which, like animals, depend on their green brethren for their livelihood.

Why arable land has to be fertilized

Let us consider Nature's eternal cycle of give-and-take. The green plant produces organic compounds in quantities beyond its own needs. The surplus is taken up by animals. In our bodies the or-ganic substances burn away and the original energy that the green plant stored reappears as heat and motion. What is left is carbon dioxide and simpler nitrogen compounds. We breathe them out and void them as urine and feces.

These waste products, useless to our body, can serve as raw materials for plants. Similarly, when after death plants and animals decompose, the simple components that emerge from this decay do likewise and thus sustain future generations of living organisms. Plants and animals over long spans of time, in their complementary needs and interdependence, have achieved a wonderful balance. This balance is maintained so long as Man does not, in his meddling way, interfere with Nature.

When we take the wheat from the fields without leaving the dying stalks to turn into humus, when we mow and carry off the grass from the meadows, instead of letting the grazing animals return in changed form to the earth what they have taken away, it is obvious that the soil will be starved and impoverished. Man realized this soon enough and started to cart the dung from his stables and cowsheds back into the fields. Thus, in the natural manure, he gave back to the soil nitrogen and other useful material. The greater the exploitation of the soil by intensive cultivation, the greater the depletion.

For several centuries potassium nitrate was imported from Chile as fertilizer. There are districts in Chile with very little rainfall, where residues of animals and plants rich in nitrogen have been deposited in enormous layers without being washed away and dissolved. Before the exhaustion of these natural sources could become a problem the chemists managed to harness the nitrogen of the air, and nowadays we have factories that specialize in producing nitrogen compounds to be used as synthetic fertilizers. With synthetic fertilizers we try to make up for our sins against Nature. But the territories where Nature is left to work out her own balance become fewer and fewer.

4. *The Function of Stomach and Intestine*

Out of sight, out of mind! Once we have eaten our meal we do not think of it any longer, except when it gives us indigestion. We are not aware that every solid meal stays for several hours in the stomach and that what goes on there and in the intestine is of more vital importance than the perfection of the cook. We do not need to think about it because as soon as we have swallowed a bite its fate is beyond will and knowledge. Carefully collected observations and many experiments on Man and animals have brought to light what is beyond the reach of our direct experience and have taught us what digestion is all about.

The way is prepared like that of the water in a garden hose. When a morsel glides without mishap across the opening of the windpipe into the right passage, which is the gullet, it goes into a saclike widening of the alimentary canal, the stomach (Fig. 16). This merges into the coiled intestine and ends with the anus, through which undigested remnants are voided. The intestine is several times the length of our body; in fact, we could wind it seven times around our hips. A worried mother has to have patience before a swallowed plum stone will reappear. It has far to go.

What happens when we digest?

A plum stone does not change on its way through the intestine. It is undigestible. Our normal food, however, is digested. That means it is changed in such a way that it can pass through the intestine wall to be taken up by the blood stream. The blood distributes the digested food throughout the body so that it can act as fuel for the work of muscles or as building material in growing tissues.

In the simplest case the food is physically dissolved in the

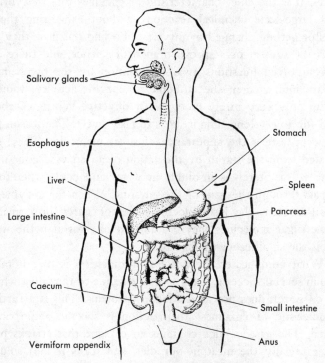

Fig. 16. The digestive system of Man.

juices of the digestive tract. The digestive fluid is secreted in special glands found in the wall of the intestine. The salivary glands produce the saliva, the wall of the stomach and the intestine the gastric and intestinal juices, the liver produces the bile and the pancreas the pancreatic juice. When we eat a lump of sugar it dissolves in the saliva as it would in a glass of water. Most of our food, however, is not soluble in water. It is the job of the digestive juices to make it soluble and to split it chemically into such small molecules that they can pass through the minute invisible pores of the intestinal wall.

The digestive juices contain, besides water, most remarkable substances, the enzymes, which attack and split the food substances. It is the chief characteristic of enzymes that they bring about large-scale chemical reactions without exhausting themselves or getting changed in any way. To find out how they do this is the subject of a special branch of science, and there are many unsolved questions awaiting solution by the biochemist. We are interested in the outcome of enzyme activity, though we can only very rarely recognize it directly. When we chew bread for a very long time it starts to taste sweet. The reason for the sweet taste is the appearance of sugar molecules. These are liberated from the starch by the action of an enzyme contained in the saliva. Starch is insoluble in water and could, therefore, not pass through the wall of the intestine. When the enzyme of the saliva turns the insoluble molecules of starch into smaller soluble sugar molecules they are able to pass through the wall. This is the significance of digestion.

What complicates the process of digestion is that it takes place in several successive stages. The enzyme of the saliva which acts on starch does not attack fats or proteins. This is a further characteristic of enzymes: their action is specific and strictly limited. The gastric juice contains an enzyme that attacks protein, especially the meat in our diet, and turns it into soluble compounds. An enzyme that digests only fat is produced by the pancreas. Often there are several enzymes involved in the digestion of each one of our basic food materials, one of the enzymes carrying the breaking down to a certain stage in the chain of reactions, while another enzyme takes over to finish the job. To go into further details is unimportant. We now understand why the activities of our digestive glands might be considered more important than those of a chef. But let us not underrate his importance all the same.

Praised be the art of cooking

To eat well-prepared food is not only one of the pleasures of life but is of immediate value to a good digestion. There is some sense in cooking meat and in not serving potatoes raw. When we prepare meat the connective tissue between the muscle fibers gets loosened and so changed by the heat that the muscle fibers become more accessible to the digestive juices. When we boil our potatoes the indigestible cellulose walls of the plant cells burst apart and liberate the nourishing starch grains, which swell in the heat and turn more digestible. Similar changes occur when we prepare vegetables or bake bread. Our mouth waters when we see and smell a well-prepared dish. This means that our salivary glands and with them, unnoticeable to us, our other digestive glands start secreting more profusely and so get ready to cope with the treat, much to our benefit and well-being.

The other side of the story

One might consider it tasteless to turn our attention from the delicious smell of a roast to less appetizing things.

If a piece of meat in the pantry is left and forgotten we notice that it goes off color and begins to smell; we say it goes bad. The causes of this decay are minute organisms, even smaller and more primitive than the unicellular plants and animals of which we talked before. As they often have the shape of little rods we call them bacteria (from the Greek *bacterion*, the rod). But not all are shaped like rods. Some are tiny spheres, and some are corkscrew-shaped (Fig. 17). Some are notorious for producing diseases, others play a most useful role in Nature's household, and a host of others are unimportant to us and go unnoticed. Here,

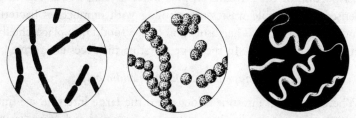

Fig. 17. Bacteria (greatly enlarged). Note three types: left, the bacilli, rod-shaped forms; the cocci, spherical; the spirilli, corkscrew-shaped forms. Bacteria of decay are generally to be found among the bacilli.

we are interested in those bacteria that bring about decay.

They are present everywhere, and without them there would be no decay at all. Every breeze carries their spores, and these are bound to settle on a piece of meat that is exposed to air. The meat offers most suitable living conditions for them. They rapidly multiply and extract from it the substances that they need for the building up of their bodies, by breaking protein into simpler chemical compounds. When, by their action, a corpse putrefies substances are liberated that serve as valuable food for plants. One peculiar characteristic of the metabolism of putrefying bacteria is that many substances are formed that smell very bad indeed. Putrefying bacteria live in great quantities in our intestine too. They take their share of food; on the other hand, in breaking down food substances, they aid the working of our digestive juices. Keeping putrefying bacteria from entering the intestines of young chicks and newborn guinea pigs has been attempted — not an easy thing to do if we consider their omnipresence. But in successful experiments the animals did not thrive or grow, and it became clear that the co-operation of putrefying bacteria is essential for digestion.

During their long journey through the intestine the useful components of our food are processed and pass in solution through the wall of the intestine. After this they may be resynthesized, sugar into glycogen (animal starch), the components of animal fats and proteins into the fat and protein of our body. In the last portion of the intestine surplus water is reabsorbed. The condensed undigestible wastes are teaming with numberless bacteria and form the feces. The absorption of the food takes place chiefly in the small intestine. In the large intestine the feces are formed.

The appendix, a nuisance and a help

Where the small intestine widens into the large intestine we find the notorious appendix. It may become inflamed and often has to be removed. Once we have lost it we do not miss it and we begin to wonder why we should have an appendix at all, if it is useless and yet a potential threat. Well, for some animals the appendix is important, and then it is much larger. Even our early ancestors may have needed it and when it lost its usefulness it became smaller. This is usually a very slow process. Nature is very conservative and drags along for hundreds of thousands of years so-called vestigial organs, vestiges of once useful structures, just because they belonged to the basic inherited outfit of the body. We shall hear more about this later on.

But what does a functional appendix do? Well, strangely enough it has to do with the fact that the walls of the plant cells are composed of cellulose, a complex carbohydrate that cannot be tackled by the digestive juices of most animals. We have already mentioned that we burst these cells apart by boiling our vegetables and so liberate the nutritive contents. The plant-eating animals solve this problem in their own fashion.

Many herbivores among the mammals, for example, the

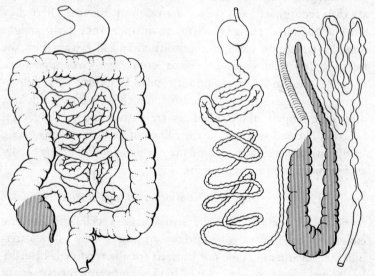

Fig. 18. Stomach and intestine, left, Man; right, rabbit. The shaded part is the caecum.

horses, guinea pigs, and rabbits, have a very long appendix (Fig. 18). It is filled with enormous quantities of bacteria that can split and dissolve cellulose. In doing this the bacteria live at the expense of their hosts but at the same time render invaluable service by dissolving the cell walls of plants and exposing their contents to the action of the digestive juices.

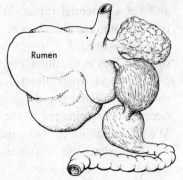

Fig. 19. Stomach of a ruminant.

We find still more extraordinary conditions in the ruminants, among the herbivores (cattle, goats, sheep, deer, etc.). Their appendix is not very long but their stomach is rather curiously designed (Fig. 19). Their food goes first into the rumen. This, like the appendix of the former group of herbivores, is a breeding place for cellulose-digesting bacteria. The contents of the rumen are then regurgitated and once more reach the mouth. There they are chewed all over again. After rumination they pass straight into the remaining parts of the stomach and afterward into the intestine for final digestion. Birds even have two appendices (Fig. 20) and in them we find a similar fraternity of cellulose-digesting bacteria. In the jungle fowl, which eat a lot of cellulose, difficult to digest, the appendices are extraordinarily large and function like the appendix or the rumen of herbivorous mammals. In the sparrow hawk, a bird of prey whose food contains hardly any cellulose, the appendices are tiny.

The lost glass beads and other strange stories

Some lower animals like the snails have a cellulose-digesting enzyme in their saliva. The intestine of others, such as the caterpillar, has neither enzyme nor bacteria for the treatment of cellulose. They have to discard a lot of food because only the content of those cells that happen to get damaged during chewing can be digested. This explains why caterpillars eat so much and do such a lot of damage.

Pages and pages could be written on the various ways of feeding. Just think of teeth, which we have so far not mentioned at all. We bite and chew with them, and it is obvious that they open up the food to the digestive juices, and once we have lost them we know how useful they were. A carp has teeth far back in its throat; a crayfish has them in its stomach. Their usefulness is the same wherever they are. Birds have no teeth at all, and yet they chew more thoroughly than we do. This was found out more than two hundred years ago by some learned gentlemen who were interested in the then still open question whether birds digest grains chemically or mechanically. They made hens swallow glass beads, the bore of which they had filled with grains. They reasoned like this: chemical digestion would dissolve the grains. If, however, the bird relied on mechanical digestion the grains should reappear. To everybody's surprise not only the

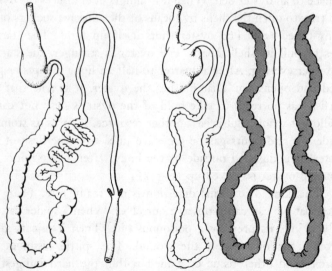

Fig. 20. Stomach and intestine. Left, the sparrow hawk; right, the grouse. The caecum is shaded in black.

Fig. 21. A starfish feeding on an oyster.

grains disappeared but likewise the beads. The stomachs of the hens had pulverized them. Their stomachs have such a tough inner lining and such strong muscles that the contents get ground up as between millstones. This is all the more effective because of the ever present particles of sand and gravel that the bird swallows with its food. An ostrich even swallows coins and nails, if there are any about, without being the worse for it but to the advantage of digestion. Thus all grain-eating birds possess in their muscular gizzard a perfect substitute for a set of teeth.

However, successful feeding is possible even without teeth or a gizzard. For instance, a starfish can eat the hard-shelled oyster! It looks so harmless and inoffensive and yet it is a perfect menace to an oyster bed. When it is hungry it embraces the oyster and tries to open the shells by means of the adhesive suckers of its many-tubed feet. The oyster shuts itself up tight. The starfish tries to pull its shells apart. The oyster is stronger, the starfish more persevering. After a quarter to half an hour the oyster gets tired and opens up. That finishes the oyster. It is true that the starfish has no teeth to take hold of the oyster's flesh nor can it swallow it whole. But it has another resource. It turns its stomach inside out and pours gastric digestive juice over the flesh of the oyster. It is digested outside of the body; after a few hours the liquefied oyster is sucked up (Fig. 21).

This kind of external digestion is not really rare. There are cases that we can easily observe ourselves. When a spider catches a fly it kills its prey by a poisonous bite. Then it injects saliva through the opening of the wound. The spittle dissolves the muscles and other tissue contained within the hard indigestible chitinous armor of the fly, and after a while the spider sucks up the now fluid contents of the fly through its narrow gullet. All that is left is the fly's empty armor, empty as the shell of an egg after breakfast.

These are just a few examples of the many feeding techniques among animals. The methods differ, but the aim is always the same: with the help of enzymes food is split up into simpler fluid components that can pass through the intestinal wall and get assimilated by the body. All the thousand and one possibilities are but variations of one main theme.

5. Body Fluids

Our body contains much more fluid than we usually realize, and some animals are even more waterlogged. If a human body weighing a hundred and forty pounds were to dry up in the desert the shriveled-up mummy would weigh about forty pounds; the remainder is water. In the human body the water is contained partly in the cells themselves, partly as tissue fluid between them, and finally in the blood. The latter is the most interesting of the body fluids. It is with its distribution in the body, its composition, and its function that we are now concerned.

Heart and blood vessels

That the heart is the seat of our emotions is a naïve conception. One speaks of winning a man's heart or of leaving him broken-hearted and of being heartless. The sober scientist, however, knows that the heart is not an organ of tenderness. It is hollow inside, has a thick muscular wall, and its task is to make the blood circulate in the body.

From birth to death the muscular wall of our heart contracts powerfully about once a second or a little faster. This is the heart-beat. It never stops. By contracting, the heart squirts forth the blood it contains in the same way as when we produce a jet of

Fig. 22. Capillary vessels: at right, in white, arterioles carry the blood to the capillaries; at left, the venules retrieve the blood (returning it to the heart). (Greatly enlarged)

water by squeezing it through our folded hands. This jet of blood runs through our body within preformed channels, the blood vessels. The vessels split into several main branches, which supply the blood to different parts of the body. The main branches subdivide again and again, become narrower and narrower, until in the end they form the finest capillaries (Fig. 22); these are much finer than hair and are invisible to the unaided eye. After the blood has passed through them, the capillaries re-form into larger vessels and the blood eventually reaches the heart again through a main vessel. The vessels that lead away from the heart are the arteries, those that lead back to it the veins. Between the smallest arteries and the smallest veins lie the capillaries. The blood never deviates from this prescribed pathway.

This may sound questionable because blood oozes out wherever we prick ourselves with a fine needle. The capillaries form such a dense network that it is quite impossible not to tear a number of them by a pinprick, and this must lead to bleeding. If one could unravel the capillaries that supply our muscles only and put them end to end they would add up to a length of 60,000 miles and reach two and a half times around the equator.

Why we do not bleed to death from a pinprick

If we bore a little hole in the bottom of a barrel the water will drip away until the barrel is empty. If we prick our finger we just lose a few drops of blood, and even when we are severely injured the bleeding eventually stops. This is most important. Because if the ten pints of blood that the body of a grown-up

person contains dripped out of
the damaged vessels like water
from the barrel, even penknives
would be fatal things to handle.
However, wounds stop bleeding
by themselves because blood can
clot. If we collect blood and let
it stand for a while it will set.
This process is somewhat similar
to the hardening of the fluid egg
white by heat. When the blood
leaves the vessels a microscopically fine network of protein
threads is formed, without any
application of heat. Contact with
air is sufficient to induce this

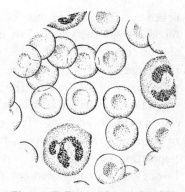

Fig. 23. Cells in a drop of blood.
The red blood cells are without
a nucleus (unlike those of the
frog, Fig. 1). The white blood
cells have a branched nucleus.
(Stained cells, microscopic view)

change. The clotted blood seals the cut vessels and bleeding stops.
Such clots can be produced even within an undamaged vessel,
especially if the inner lining has deteriorated with age. The blood
stream may deposit a clot in an important vessel and thus cause
a blockage leading to a stroke. On the other hand there are persons whose blood does not clot at all, owing to a peculiar hereditary disposition. These bleeders lead a very precarious existence,
as the slightest injury can cause a critical loss of blood.

Why can the body not stand an excessive loss of blood? And
why can an organ, cut off from its normal blood supply, not
remain alive?

A drop of blood under the microscope

In order to live, the cells of our body need oxygen. The presence
of oxygen is essential for the proper functioning of the metabolic
processes in the protoplasm. It is one of the most important tasks
of the blood to supply the cells with oxygen. Let us first look at
a drop of blood under the microscope.

It is no longer the uniformly red fluid it appeared to the

naked eye. Instead we notice countless reddish round cells, the red blood corpuscles, which float in a yellowish or, in a thin layer, almost colorless fluid. Besides, one notices a few colorless cells, the white blood corpuscles, of which we see one complete, and two in part, in the drawing (Fig. 23).

The red cells are so small that 1,250 of them end to end in a row would be just two-fifths of an inch long, and the number of them in a fair-sized drop of blood roughly equals that of the population of the United States. They contain a red pigment, hemoglobin. Hemoglobin is a protein capable of absorbing oxygen easily and releasing it just as easily into an environment of low oxygen concentration. When it carries oxygen it is a light bright red; it is dark red after it has given it up. The color is probably of no functional importance. The blood of a cuttlefish contains a related substance doing the same job, but it is blue when it carries oxygen and colorless otherwise. In order to understand the working of the oxygen carrier let us study the blood circulation a little more closely.

The circulating blood and its function

Our heart is divided into a right and a left half by a longitudinal wall. With each heartbeat the blood in the right half (drawn black in the figure; the human being is seen in front view; the right half of the body comes to lie left in the figure) is pumped into the lungs through the pulmonary artery, while at the same time the blood in the left half (white) is pumped into the whole body through the aortic vessel (Fig. 24). During the relaxation of the heart muscles both halves fill up again with blood coming in through the big veins.

The red blood cells thus take the following route: coming out from the right side of the heart, they eventually reach the blood capillaries of the lungs. These have such thin walls that the oxygen, which we breathe in with the air, easily diffuses into the blood where the hemoglobin grabs hold of it. Then the red oxygen-laden cells return via the pulmonary vein to the left half of the heart, whence they reach the tissues of the body where the oxygen diffuses through capillary walls into the adjoining cells.

The rhythm of this circulation is brisk. Within less than half a minute the blood cells complete the described journey and they thus transport great quantities of oxygen. We know already that chemical substances are slowly oxidized and that the end products of this oxidation are water (H_2O), carbon dioxide (CO_2), and in addition ammonia (NH_3) if nitrogen-containing proteins are involved. Ammonia is poisonous and is rapidly transformed into somewhat less harmful compounds such as urea.

A fire in a grate needs fresh air (that is, its oxygen), and the carbon dioxide and water, the gaseous end products of combustion, leave through the chimney. No flames or smoke arise during slow oxidation in the cell. But the end products of chemical changes in the cell have to be removed. While the blood coming into the tissues brings the oxygen, on leaving them it acts like the chimney by carrying away the carbon dioxide and voiding it into the air through the thin capillary walls of the lungs, which it reaches via the right half of the heart.

Urea too is removed by the blood stream. Since it is not a gaseous substance it cannot pass

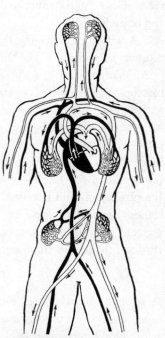

Fig. 24. Diagram of the circulation within Man. Black, venous blood; white, arterial blood. (Except for the pulmonary circulation, where arteries carry blood that is not oxygenated, the pulmonary veins, returning to the heart, carry oxygenated blood.)

through the lungs but is filtered through a special excretory organ, the kidney. The urea, together with other mineral waste products in aqueous solution, makes up the urine.

We might find it surprising that the end products of oxidation in our body can be removed so easily by the lungs and the kidneys, while in our grate we find after a roaring fire masses of ashes and clinkers. But then ashes and clinkers are not burned material but remnants to be compared to the undigestible contents of our food. These are separated off in the gut and leave it as feces. Only high-quality oxidizable substances reach the blood and body fluids.

Apart from managing the gas exchange in the tissues and the removal of waste material the blood is concerned with fuel transport. We know that a fire goes out if we do not stoke it. The metabolism in our tissues would stop if we did not refuel, and it is the fluid food-laden blood plasma that seeps through the submicroscopically small pores of the capillary walls into the intercellular spaces of the tissues that delivers the food to the cells. Since the red blood cells are too large to squeeze through the pores, the tissue fluid is not red but clear and yellowish. The lymph, as this fluid is called, after having given off the food substances serving as fuel, or for the building up of new living matter, is collected in special lymphatic vessels and led back through them into the blood stream.

The police

A drop of blood under the microscope showed us besides the countless red corpuscles a smaller number of colorless ones, the white blood corpuscles (Fig. 23). Both have in common that they do not form a coherent tissue but are carried about singly by the blood stream. However, while the red cells get swept along like so many lifeless specks of dust to wherever the blood happens to flow, the white cells show a certain independence. Like little amoebae, they can produce protoplasmic appendages. With these pseudopodia they can stick to the inner walls of blood vessels, and they can pass through the finest pores of the capillary cells and then wander about between the tissue cells. Furthermore, they can feed like amoebae and, with their pseudopodia, surround and take up solid objects smaller than themselves. Two of their many different tasks are quite well understood: they engulf intruding bacteria, and they remove damaged tissue cells.

We mentioned earlier the putrefying bacteria. Of the many different kinds of bacteria some are dangerous and cause diseases. If a wound from a rusty nail gets inflamed, this is not because of the rust, which itself is quite harmless, but because certain bacteria that either were on the nail or stuck to the lacerated skin got into the wound, found the living conditions favorable and started multiplying. It is their poisonous metabolites that damage the tissue.

The redness of an inflamed piece of skin is caused by an intensified flow of blood which brings along a greater number of blood corpuscles. The white blood cells accumulate in the inflamed tissue and fight the bacteria by eating them up and digesting them. The more quickly they act the sooner they overwhelm the bacteria and the inflammation subsides. If the bacteria win the race and get a firm hold, they multiply, damage the tissue by destroying the cells, then more and more white corpuscles are called in for the defense, so many, in fact, that the body fluid at this patch turns whitish and pus is formed. The whole assembly of blood cells has plenty to do. Not only have they to fight the bacteria, but they have to remove the destroyed cells and tidy up. If the bacteria win and spread from the original focus all over the body and swamp it with their poison, we suffer from blood poisoning.

Immunity and immunization

Each of the notorious infectious diseases such as tuberculosis, typhus, dysentery, cholera, plague, diphtheria, and so on is caused by a specific kind of bacterium. Not every bacterial invasion is successful in starting disease. The white blood cells are the first line of defense and try to wipe out the invaders. In addition the blood plasma contains substances that conduct a kind of chemical warfare. Our natural resistance (natural immunity) against infections depends on the quantity and effectiveness of these defense forces. We owe it to them that not every swallowed typhus bacillus brings on typhus and that we city dwellers do not all suffer from tuberculosis, in spite of breathing in tuberculosis germs with the street dust.

We are not left completely powerless even if the disease-carrying bacteria succeed in multiplying and spreading in our body. Antibodies are formed, partly against the bacteria directly, partly against the powerful poisons produced by them. The efficient and speedy formation of antibodies in contest with the vitality and resistance of the bacteria will decide the outcome of the battle and whether we shall live or die.

Where and how antibodies are formed is not exactly known. A few days after the onset of an illness their presence in the blood fluid can be proved. They are effective only against those bacteria and their toxins that caused their formation. Antibodies produced against typhus bacteria can fight typhus bacteria only and are powerless against any other disease-carrying organisms. Antibodies against smallpox only fight smallpox. Strict specificity is one of their features. When the body has recovered from an attack they can prevent reinfection by the same bacteria for a long time afterward. Against some diseases like smallpox and scarlet fever we acquire a lifelong immunity.

These discoveries have led to far-reaching applications. If, for instance, one injects pathogenic bacteria into a horse it will form, in its blood stream, antibodies against them. If we then remove the blood cells from the blood we can use the healing serum for immunization. One can immunize human beings against a disease by injecting them with an animal serum containing the specific antibodies, and if the disease has caught us unprepared we can speed up recovery by doing so. In other cases, for instance with smallpox, one injects the bacteria themselves, or rather a

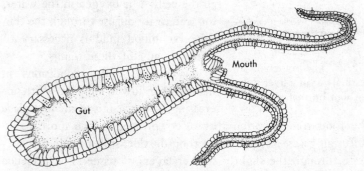

Fig. 25. Longitudinal section through the Hydra. (Greatly enlarged)

weakened and therefore harmless strain, which stimulates the human body actively to produce the antibodies, which then prevent any infection for years to come.

Some people are antivaccinationists. If they knew how much suffering has been prevented since vaccination has been introduced they would quickly be converted. Since the introduction of preventive vaccination the disease has practically disappeared in most countries that practice modern medicine, while in other countries that do not have compulsory vaccination, and in spite of good hygienic conditions, this scourge still claims its victims.

Heartless animals, and such as have a heart in every leg

Having talked so much about the importance of the blood, I hardly dare say it: we know of animals that have neither a heart nor blood vessels. Are these organs not essential for living, after all? Do not let us draw hasty conclusions; all depends on the build and size of an animal.

In ponds and ditches there lives a fresh-water polyp, the Hydra, that is less than a sixteenth of an inch long (Fig. 25). Its body wall consists of only two cell layers which enclose a big gut. Around the mouth opening waft tentacles that see to it that the animal is properly filled with food. Owing to the thinness of the body wall, what the gut digests need not travel far to reach

Fig. 26. The branching intestinal gut of a liver fluke. (About normal size)

all the cells. The oxygen in the water, too, can easily diffuse through the thin skin. No blood fluid is necessary to transport any of these things.

We find similar conditions in many other lower animals. With increasing body size it is in the first place the food transportation that offers problems. The more easily diffusing gases, oxygen and carbon dioxide, still manage to penetrate through the slightly thicker layers of tissue. The presence of a branched gut (Fig. 26), which spreads through the whole body, is one way out of this difficulty. Thus the food is brought near the cells in all parts of the body. Removal of the more cumbersome metabolic waste products is accomplished by renal organs, which also branch throughout the body and collect waste from everywhere.

When the body grows still bigger, then even lower animals need blood, blood vessels, and a heart. Their circulatory system is usually less perfect than in Man. Often capillaries are missing, often the arteries are short and stop altogether so that the blood pours freely into the intercellular tissue spaces, whence it is taken up by short veins and sent back to the heart. Certain crustacea and insects have no blood vessels at all but a tube or saclike heart which by constant beating stirs the body fluids like a mixer — a method sufficient for the unpretentious.

However, the long, thin legs of some insects, such as the plant-sucking aphids, would be bypassed and suffer from lack of blood were it not for each leg having its own small, private heart to pump in its share of the precious fluid.

All vertebrates, however, beginning with the cold-blooded fish, have a closed blood circulation, similar to that of Man. The number of blood capillaries and the density of the network of vessels are smaller in a fish than in a bird or mammal. The latter are warm-blooded, their metabolism is brisk, and they need a more efficient blood circulation.

6. *Respiration*

Why does Man suffocate in water and a fish in air?

Many an owner of an aquarium will tell you that one of his more lively pets jumped out and was found dried up on the floor in the morning. In actual fact it had suffocated, though it does sound strange that it should have died from lack of air. The reason is that the respiratory organ of a fish is quite different from that of a land animal. What actually is a respiratory organ?

We know that living cells depend on oxygen. We have heard how this is supplied by the blood. Cells therefore are said to "breathe." Generally, however, one understands by breathing the traffic of air through the lungs, where the blood is loaded with oxygen to be conveyed to the tissues. Organs whose job it is to procure oxygen from the environment to hand it on to a body fluid are called respiratory organs.

The lungs and their ventilation

The lungs of a salamander are of simple design. They are two thin-walled air-filled sacs that are ventilated via the nostrils, the throat, and the windpipe. In the walls of the lungs the branches of the pulmonary artery end in capillaries. The gas exchange between the air in the lungs and the blood in the lung walls is the more vigorous the larger the available respiratory surface. This holds for any kind of exchange on surfaces. We make use of this principle when we pour hot milk from a cup into the saucer in order to cool it more quickly. Or another example that brings us a bit nearer to our case: if one keeps fish in a round goldfish bowl one can see the poor creatures come frequently to the surface for air, because the water surface that is in contact

with the air is much too small
for sufficient air to enter the
water and to replenish the used-
up oxygen. In an oblong aquarium, however, this surface is
much larger and the fish are not
starved for air, provided we do
not keep too many in one tank.

In the salamander the inner
surface of the bladderlike lung
is sufficient for the gas exchange.
Its metabolism is as slow as the
animal itself and not much oxygen is needed. Things are quite
different in the lively lizard. Here
the walls of the lungs show

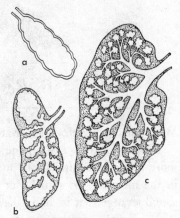

Fig. 27. Section through lungs
of (a) salamander, (b) lizard,
(c) Man.

numerous crevices, and thus we now get hundreds of little
bubblelike structures, called alveoli, through the walls of which
blood can circulate (Fig. 27). The human lung is built on the
same principle except that there is no common air space left, as
the whole interior is subdivided into alveoli. This brings about an
enormous surface enlargement, and it is obvious that in this way
the gas exchange between blood and air in the lung is extremely
efficient.

Naturally the used-up air has to be replaced. This is brought
about by breathing. If we pull on the plunger of a syringe, we
draw in air through the opening. We cannot see the air. But when
we do the same, this time with water, we can see it rise in the
syringe.

Our lungs are ventilated in a similar way. Their free volume
corresponds to the volume of the syringe, its opening to our
nostrils, whence the path leads via the nasal cavity, throat, larynx,
and windpipe to the lung cavity. The plunger is our diaphragm, a
dome-shaped muscle that separates the chest from the abdomen.
When we breathe in, the diaphragm contracts (shortens) and the

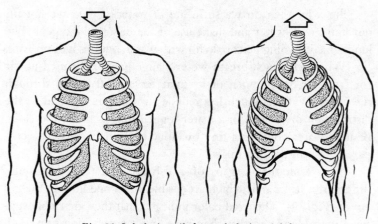

Fig. 28. Inhalation (left), exhalation (right).

dome flattens (Fig. 28), the plunger moves downward, the volume of the thorax enlarges, and air is drawn in. At the same time certain muscles contract and lift the ribs, thus enlarging the thorax sideways. When the muscles relax the elastic walls of the lungs contract and push the air out. Exhaling can be accelerated by a reverse movement of the ribs and by the contraction of the abdominal muscles. We can never empty our lungs completely. With every breath we take, only a certain portion of the air is replaced: the lungs are ventilated.

Why Man cannot breathe under water and fish suffocate in the air

We have already mentioned that water which is in contact with air takes up oxygen. In fact all the gases of which air is composed are to a certain amount soluble in water. One can see them just as little as dissolved sugar. But if we heat water, gas bubbles rise because hot water can dissolve less gas than cold. If water did not take up oxygen our lungs would be of no use to us. Only the oxygen that has entered the watery fluid in the moist lung cavity and then the blood fluid in the capillaries can be absorbed by the red corpuscles and transported to the tissues.

But why then do we suffocate in water? Could we not fill our lungs with water and make use of the dissolved oxygen? The lungs of a drowning person do fill with water, but this is of no avail.

When with each breath we exchange part of the used-up air for fresh air, the oxygen of its own accord diffuses all through the lungs. Water is much less mobile than gaseous air and so the distribution of oxygen in a water-logged lung is much too slow. Besides, water contains a tiny quantity of available oxygen compared with the amount in air.

The respiratory organs of a fish are therefore built quite differently. It has gills which are visible when one lifts up the gill cover (Fig. 29). Their red color indicates that they are richly supplied with blood and that they have an extremely thin skin, both factors being favorable to an efficient exchange of gas with the surroundings. In contrast to the lungs, we have in this case an enlargement of the outer surfaces in the development of little gill lamellae. Only when we look at them under the microscope do we realize the extent of the surface area. Invisible to the naked eye, they carry on both sides many thousands of tender, projecting ridges into which capillaries branch out. The water, which a breathing fish continuously pumps past its gills, meets these folded lamellae and the dissolved oxygen enters through them.

If the fish is thrown onto dry land its gill lamellae, which in the water open up and spread out, now all stick together. The respiratory surface is reduced to a fraction of its functioning size. It is of no use that the air contains plenty of oxygen; the fish suffocates long before it dries up.

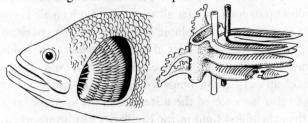

Fig. 29. Gills of the fish. Right, three filaments. (Greatly magnified)

So we see that respiratory organs are highly adapted to special types of surroundings. To Man in water, however, and to the fish on land their very perfection spells disaster.

Oxygen is needed by the cells of lower animals too. Yet small and fragile creatures often have no respiratory organs because their gas exchange takes place through the outer surface of their skin. Plants breathe through tiny openings in their covering layer of cells. During the metabolic processes in plants oxygen is absorbed and highly complex chemical substances are split up and broken down; in other words oxidation takes place in essentially the same way as in animal cells. The living plant too needs a supply of oxygen. No organs corresponding to lungs or gills are needed because the body of a plant branches considerably and as a rule develops a larger surface area through which sufficient oxygen can be taken in. But if plants have no lungs, they have something resembling nostrils, very small, microscopically small, nostrils. Called stomata, they sit on the leaves of plants and form a passage through which the gas exchange between the surrounding air and the air spaces inside the leaves is regulated. They are so numerous that on the underside of a leaf of a peach tree from one hundred to two hundred are distributed over an area of 1/2,000 of a square inch (Fig. 30).

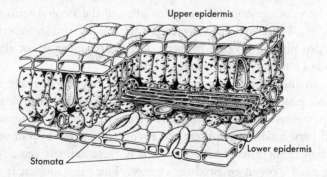

Upper epidermis

Lower epidermis

Stomata

Fig. 30. A cross section through a plant leaf. On the lower epidermis, the stomatal openings. (Greatly enlarged)

In all green plants the oxygen consumption can most easily be shown at night, when the plant is in the dark. We have learned that in daylight the plant is busy synthesizing, with the help of chlorophyll, organic substances from carbon dioxide and water, while oxygen is liberated and given off into the air as a by-product. At the same time the tissues of the plant respire continuously and the cells use up oxygen. However, during the day photosynthesis is the dominant reaction, and a surplus of oxygen results. All this explains why the oxygen content of the air does not diminish, although it is used up constantly by innumerable animals and in a more modest way even in plants.

Life without oxygen

It is by no means a law of Nature that all living creatures need to breathe. There exist organisms that can live without oxygen. This means they derive their energy not from oxidation processes but from chemical reactions of a different kind. The yeast cells, small unicellular plants, split sugar by means of an enzyme into carbon dioxide and alcohol. During this reaction energy is liberated, which sustains life in the yeast cells. When we brew beer, wine, or other spirits we make use of the yeast by adding it to the sugary fluids. The sugar is broken down by the enzyme of the yeast: gaseous CO_2 is given off, and the alcohol remains as residue. There are many bacteria that live by similar reactions, oxygen playing no part in them. The end product is not always alcohol but may be something different.

In the animal kingdom life without oxygen is fairly rare. There exist a few oddities that can do without this vital element. Tapeworms and roundworms, which are parasites in the intestine, find very little oxygen there and even that gets used up in putrefaction. Experiments have shown that these animals actually do not need oxygen to produce energy. Their metabolism is altogether a luxury metabolism in which still valuable substances, like valeric acid, are wastefully excreted. But only he who spends

more than he owns is a wastrel. A tapeworm in the intestine
lives in such affluence that it does not need to be thrifty with food.

7. *Body Temperature*

Many people wear a fur coat "for warmth" in winter, but few
will give it a thought that the fur in itself is not really warm at
all. In fact it has the same temperature as its environment. It
does not warm us but it keeps us warm.

Why we need a winter coat and why the sparrows and lizards do not

The source of heat is our body, not the fur. The heat is produced
during slow oxidation processes that we call cell respiration. The
fur is specially suited to preserve the heat and to prevent it from
flowing off into the cold surroundings. Nobody would think of
keeping the cold air out by wearing a coat made of tin sheet or
for that matter a knight's armor. Metal is a good conductor of
heat and therefore a bad protector against cold.

When we put a frying pan on the fire we know that its
iron handle very quickly gets too hot to be touched, because
iron conducts heat so well. We prefer our frying pans to have
wooden handles since wood is a poor heat conductor. Water is
a better conductor of heat than air; therefore we find a room of
68 degrees quite comfortable even if we wear just a bathing
suit, while bath water of the same temperature is unpleasantly
cool. The water drains our warm body of its heat much more
quickly than the air. Another poor conductor of heat is the horn
of our fingernails, and this can be easily proved. An object that
is too hot to be touched for more than a split second with our
hands or lips can be in contact with a fingernail for quite a time

before the heat reaches the sensitive layer under the nail. Hair and feathers consist of the same horny substance as nails. They, too, are poor conductors of heat and a very good protection against cold. This effect is intensified by the layer of air caught between them.

This is the reason why a sparrow and a goose, a hare and even a bear in Siberia do not need a special winter coat. Nature has planted a fur coat right into their very skins. Man, having only a sparse remnant of a natural hair coat, has hunted animals for their fur since time immemorial, and we still use hair blankets and eiderdowns for cover.

Only mammals and birds have protective hair or feathers. All the other vertebrates, like lizards, frogs, and fish, and all invertebrates, like crabs, snails, and worms, have a naked skin. The body temperature of a lizard changes with the temperature of the environment. There exists a strange correlation: in hot sunshine the lizard is lively, during a cool spell it gets lazy, and during cold nights or in winter it will be rigid. Although the animal produces heat by oxidation in its body it cannot retain it and is thus to a high degree dependent on the temperature of its environment. All reptiles, amphibians, and fish, all insects and the lower animals change their temperature. One calls them cold-blooded, but this does not quite meet the case, because in the hot sun the blood of a lizard reaches a higher temperature than that of a human being. They are variably warm.

Birds and mammals are warm-blooded. They keep approximately the same temperature during summer and winter, by day and night, in sunshine and in rain. This is good, because it has made their metabolism and all their bodily functions independent of the environment. However, safeguards are needed to prevent a loss of body temperature to the surroundings, and one of the most important ones is a hair or feather coat. We, with a body temperature of approximately 98° F., have to wear clothes in order to maintain this temperature in all kinds of weather.

Prevention of heat loss alone does not explain how we can

keep up a steady body temperature when the temperature of the environment changes. Special methods are employed to compensate for loss or gain of heat.

The regulation of the body temperature

When it gets cool we put on warmer clothes. Like our clothes, the thicker hair coat grown by animals in winter and lost again in spring can serve as a coarse temperature adjustment only.

The continuously necessary delicate adjustment is of a different kind. It functions independently of our will and often without our knowledge, whenever the temperature changes from the 98° F. to which the working of our cells is geared. When our body gets too warm the blood vessels in the skin dilate, our face gets red, and we give off heat. At the same time the oxidative processes in our body are cut down to a minimum and less heat is produced as we stop all vigorous movement. Moreover, our sweat glands become active, we start sweating profusely, and heat is used up to evaporate the perspiration. This also cools the body. Dogs, which have no sweat glands, hang out their tongue instead and pant vigorously, using their tongue and lungs for cooling.

When on the other hand our body cools down, the blood is withdrawn from the skin into the internal organs reducing heat loss of the blood, food combustion increases, and if this is not enough we shiver with cold. This means that our muscles move involuntarily and thus produce heat.

Summer and winter this regulating mechanism is so accurate that our temperature rises only during a fever when an illness upsets the balance, or it falls a few degrees in a state of exhaustion. Some mammals have a slightly lower, most a slightly higher, body temperature than we. Birds generally have a normal temperature of up to 108° F., which Man never reaches except in a fatal attack of fever.

On the other hand, social insects cannot keep their own body temperature independent of their environment, but they do

protect their most treasured possession, their brood, from overheating or from cooling down. On a hot day one can see field
wasps flying busily to and fro between their nest and a nearby
water puddle. The nest of the field wasps is an open one, usually
built on stones, beams, or branches and exposed to weather
intemperances. In their stomachs they fetch water and spit it over
the paperlike substance of which their nest is composed. Then
they sit down on it and, as living ventilators, start fanning vigorously with their wings. The water quickly evaporates and thus
cools the nest. When, on the other hand, it gets cold they cluster
over the nest and try to reduce the loss of heat by covering it
with their bodies.

We find the most perfect thermo-regulation in the beehive,
where the brood cells are kept day and night at a constant
temperature of 95° F. Bees, like the field wasps, carry water,
spread it in countless little puddles over the comb, and fan it
when things get too hot. Quick and harmonious collaboration
by many individuals makes this air-conditioning the more effective. When the temperature falls they heat the nest by crowding
by the thousands into dense clusters all over the comb. Bees do
not belong to the warm-blooded animals but they can raise their
body temperature about 20° F. above the temperature of the
surroundings by speeding up the metabolic processes, and though
this may not be noticeable in a single bee in the open air, it will
be more effective when thousands of these little living stoves sit
together in an enclosed beehive.

Under similar conditions other creatures too and even plants
do generate heat. When, for malting, a pile of barley grains is
brought to germination, the temperature close to the pile may
rise by 10 to 20° F. above the room temperature. In plants the
development of heat is usually masked by heat-consuming reactions such as photosynthesis. In the interior of flowers, however,
the rise of temperature due to respiration may become measurably
higher. Insects, which often make the calyx of a flower their

nightly quarters, find there not only protection but a warm, cozy little room. Many an alpine plant that flowers at the border of a snow field uses its warm breath to melt its way through hard crusts of ice and thus reaches the light and the warmth of the sun.

8. *The Sense Organs as a Bridge between the World and Experience*

To the sense organs we owe all knowledge about the world around us. From the very first day of life every kind of experience has reached us through the sense organs. They are the basis for all perception, imagination and activity as well as for our ideas about the world, whether we are simple-minded folk or philosophers.

A journey of exploration of our own body

We commonly speak of our five senses, those of sight, hearing, smell, taste, and touch. But a human being with only these five senses would be a freak, and we can easily find out that this list is incomplete if we care to explore our own body. Let us choose a 1½ square inch of skin on our lower arm and draw an ink line around it. If we press gently with our finger on this piece of skin we have the sensation of touch. If, however, we touch this same piece of skin with the end of a very fine bristle, we shall make the extraordinary observation that only certain points are sensitive to touch while the rest are not. If we mark these points with ink and count their number in the square patch of skin we shall find roughly 120 such touch-sensitive spots. They are equipped with nerve endings and represent the touch receptors of the skin. Their protoplasm gets excited by the slightest pres-

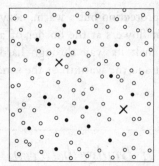

Fig. 31. Sensory spots on the skin of the forearm (about 1½ square inches). ● — Sensory spots for "cold stimuli." X — Sensory spots for "heat stimuli."

sure and this excitation is conducted with great speed by the nerve fiber to the brain. There the nerve cells of the brain translate this stimulus for us into a conscious sensation, which we can trace to and feel at the touched spot. If for stimulation we use still finer bristles or a hair we shall find that the touch stimulus needs to be of a certain strength or intensity in order to evoke a response. The same holds for vision and hearing as well as for any other sensory function. The intensity of the stimulus that is necessary to evoke a just perceptible sensation is called the threshold value of the stimulus. This expression pictures stimuli as having a certain intensity which has to be great enough to take them over the threshold of the door that leads to consciousness. A low threshold means great sensitivity, a high threshold poor sensitivity.

Now let us continue our exploration. If we touch a piece of skin with a hot or a cold needle we find that the touch spots are insensitive to temperature. Temperature receptors lie in between the tactile ones (Fig. 31). More surprising still, heat and cold stimuli are perceived in separate places. If we paint red dots on heat receptors and blue dots at the spots sensitive to cold, we find first of all that their position is a fixed one and secondly that our 1½ square inch of skin contains about twenty receptors for cold and two for heat. More densely packed than the touch and hot and cold receptors are the pain receptors in the skin. They too are microscopically small, in Man about 650 to the 1½ square inch, and they can be separately stimulated with a very fine needle point. You might consider the sensation of pain as a most unwelcome and therefore superfluous one. But you would be wrong. In fact we could not survive without its life-saving help.

If touching a flame or getting wounded did not hurt intensely, we would already have perished in childhood from carelessly contracted wounds.

As we have seen, there are more than just five senses. It would lead us too far afield to enumerate them all. Some of them we have seen to be distributed inconspicuously all over the skin, while others like the ear and the eye not only occupy a conspicuous place but show special equipment such as the eyeball and ear which makes them recognizable and familiar to everybody. The most important parts of a sense organ are the sensory cells. They respond in accordance with the function of a given sense organ to either light, sound, or touch stimuli.

There are more things between heaven and earth . . .

We must not consider our sense organs to be mere doors through which the events in the outside world can enter directly provided they give the right password. Light, sound, heat, and touch do not exist in the way we experience them. If they did, stimuli that managed to reach our consciousness through an inappropriate sense organ would nevertheless arrive unchanged. This is, however, not so. A tactile stimulus does not excite our eye because it does not reach the light-sensitive cells. Only a blow on the eye excites them, and then we do not perceive pressure but a flash of light. A heat spot in the skin does not respond to moderate pressure, but a very strong push might give rise to a sensation of heat. The nature of the sensation does not depend on the type of stimulus but on the sense organ through which it is conveyed. Given an appropriate stimulus, each sense organ can, within its allotted range of function, be relied upon to give fairly accurate information. Even then, however, sense organs are not always absolutely reliable, and this can be easily demonstrated. Our eyes seem to prove convincingly that the four vertical lines in the diagram (Fig. 32) are not parallel. That they are in fact so can be easily shown by means of a ruler. We have, therefore, been victims of an optical illusion. If we put, for some minutes,

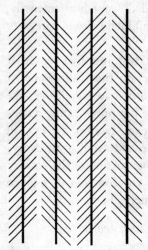

Fig. 32. Optical illusion. Are the vertical lines parallel, or are they not? They are.

our right hand into cold and our left hand into hot water and then put both hands into a bucket with lukewarm water, it will feel warm to the right and cold to the left hand, because they have been stimulated differently before. We find that our sensations are not only directly dependent on the stimuli to which we relate them but that they can be influenced to a high degree by preceding stimulation.

There are other reasons why our sense organs cannot be reliable guides to reality. To many phenomena in the outer world they do not react at all. Our eye, for example, responds to light of a wave length of between 4000 and 8000 angstroms. One angstrom is about 1/250,000,000,000 of an inch. There are wave lengths that are shorter and others that are longer than the ones just mentioned. We have no organ for the perception of either. The electromagnetic waves of our radio transmittors traverse space in all directions. They reach our body, yet we become aware of them only with the help of a radio set, which makes them audible by changing them into sound. Our sense organs select from among the many stimuli representing the outer world. This selection is not arbitrary but dictated by the way they are designed. Thus not even all of us, and certainly not all animals, may select equally. There are animals without eyes and there are some without ears. There are animals with sense organs that we do not have and that perceive things unknown to us.

For an oyster, without eyes or ears, grown onto a rock in the sea, the world ends where its body ends. No sense organ connects it with the outer world. No sun, moon, or stars exist for it, although their light may well penetrate down to the oyster bed. And when we look up at the night sky and see the stars and nebulae and their light, this fastest runner in the universe which may yet have taken several hundred thousand years to reach us here on earth, we may feel very superior to the oyster and not stop to consider how poor and limited we ourselves might appear to a still more highly developed being. The limitations of our senses are the boundary of our comprehension. Only Nature seems to be without limits.

9. *Smell and Taste*

After a visit to an art exhibition or after a play we are likely to talk about the beautiful things we have seen or heard. When we return from a mountain-climbing trip, we bubble over with tales of views and beauty. Very rarely would we speak of "beautiful smells," and a person returning from an excursion with no other news than that he enjoyed the taste of what he had eaten on the road would be looked upon by his friends with mild surprise. This shows how little attention civilized people pay to smell and taste as compared with sight and sound. The eye and the ear are Man's foremost sense organs. It would be wrong to assume that this holds for animals too.

The senses of taste and smell have in common that they respond to chemical stimuli.

The sense of smell is located in the cavity of the nose. The upper region of the nose is covered by a membrane that is kept moist. Under the microscope this membrane can be seen to consist of numerous sensory cells (Fig. 33) from which nerve fibers lead to the brain.

It is typical of all substances that yield an odor that they evaporate. A moth ball in our wardrobe gets distinctly smaller within a few weeks and eventually disappears. On a sensitive balance we can demonstrate a loss of weight within a few minutes. Molecules continuously leave the surface as vapor and move off. They enter our nose with the air we breathe and stick to the moist nasal membrane. It is these molecules that bring about a sensation of smell. The nature of the scent depends on the chemical composition of a substance, and there are so many different scents that our language lacks the means to describe them. We therefore name the scents after their source and speak of the scent of lilac and lavender or we say it smells of burned milk or spoiled meat.

For many scents we have a very low threshold. This means that our sense of smell can be very keen. Chemists often make use of this as one can still smell minute traces of some substances, the presence of which can no longer be shown even by the most sensitive chemical reactions. Artificial musk, for example, can still be spotted when we take in with one breath 0.000,000,000,003 grams of it. This is about 2,000,000,000 molecules although in fact several million times less than the smallest visible amount of it.

The sense of taste is located on the surface of the tongue. There the sensory cells lie arranged in little groups (taste buds) on small projections called papillae and partly in the wall of ring-shaped crevices, where the tasting substances get a good chance to make efficient contact with the sensory cells. These are stimulated only by substances dissolved in the saliva.

The sense of taste is not as keen nor does it need to be as acute as the sense of smell. Its task is to check the chemical quality of the food within the mouth, while the sense of smell by responding to the smallest quantities reaches far afield into the outside world.

The sense of taste is inferior to that of smell in yet another aspect. Roughly we distinguish only four qualities of taste: sweet, sour, bitter, salty. However, when we eat a well-cooked meal

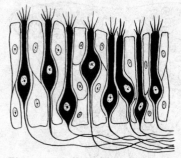

Fig. 33. Cells of the mucous layer of the nose (greatly enlarged). White, mucous cells; black, olfactory sensory cells with their nerve fibers.

or sip a fragrant drink, we notice quite a few other things besides; this is not due to taste but to the volatile scents emanating from the food and drink, which reach our nose through a passage behind the palate, at the back of our mouth. This explains why, when our nose is blocked by a cold, our food seems tasteless. Usually we are not aware what an important contribution our sense of smell makes to the taste of things.

Through practice we can heighten the efficiency of our sense of smell quite considerably. This is shown by people who use their nose in their job. The winetaster recognizes not only the local origin of a wine but even its vintage by its "taste." Actually, he judges by the volatile scents of the wine. It is unbelievable what the tobacco expert or the teataster can tell from the fragrance of his wares. And yet in comparison to many animals his performance is rudimentary.

Master noses

The design of the nasal cavity of most mammals tells us that their sense of smell must be more acute than our own. While the sensory layer in our own nose forms a small patch on the uppermost nasal labyrinth (shown black in the diagram) and the dividing wall of the nose, that of the deer occupies more than half the nasal cavity. The true size of the sensory surface, however, is shown when we look at a cross section of the inner nose (Fig. 34). In a deer the complicated folds between the conchs represent a maze of air-filled passages, which are all covered with cells sensitive to scent, whereas in Man each nasal cavity contains only the one pocket covered by a sensory lining (Fig. 35). Most other mammals have a nasal cavity similar to that of the deer, which explains their excellent powers of spooring. We know that dogs perceive certain scents in dilutions a million times below our own threshold.

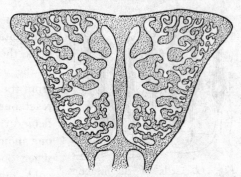

Fig. 34. Section through nasal passages of deer.

To wild animals it is important to find and to follow the spoors of their own kind. Nature has supplied them with special aids toward this end. The deer has between the hoofs of its hind legs a scent gland, the secretion of which is added to the spoor. As everything has its two sides, their enemies, the carnivorous animals, know this scent too and use it for their own benefit.

In most mammals the leading sense is the sense of smell, the eye being of secondary importance. If they showed creative instincts and had acquired culture and art they would hardly paint but might compose scent symphonies.

It is easy to see why Man, with his upright gait, no longer uses his nose to follow spoors on the ground; neither do the apes, which live in trees, nor the bats and birds, which have conquered the air. They all have a feebly developed sense of smell in comparison to animals with their noses closer to the ground.

The sense of smell in insects — a strange love story

In insects we find the organs of smell in rather unexpected places. They have no nose. The openings through which they breathe lie chiefly on both sides of the abdomen. The receptors for smell, however, are more conveniently located on the head, that part of the body that explores the things ahead.

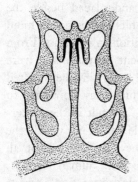

Fig. 35. Section through nasal passages of Man.

In most insects the feelers carry the scent receptors. In the bees, their performance has been very thoroughly studied, and we know that it corresponds closely to that of our nose. There are, however, some insects with a fascinatingly superior sense of smell. Some wasps with a fantastically long sting can smell out the location of a wood beetle larva living deep down in the trunk of a healthy tree, where they prick it and deposit an egg into it. This is their way of looking after their brood. The offspring of the wasp that hatches from the egg so deposited is a whitish little larva (maggot). It eats up the living woodworm from inside, then it forms a pupa, the way caterpillars do, and eventually changes into the winged wasp which in its turn uses its feelers to find yet another generation of woodworms.

Among some moths only the males have an incredibly keen sense of smell. In mammals we guessed their high sensitivity to smell from the complicated shape of their inner nose. The same holds true for the antennae of the male moth (Fig. 36). The males depend on their acute sense of smell because this is how they find the females. The attractive scent of the female is given off by glands on its body. These inconspicuous structures have been

Fig. 36. Left, a male, right, female of a moth. Note the antennae of each.

Fig. 37. Gurnard, with movable spines.

removed and placed beside the female, which seems not to mind this operation in the least. From this moment on the males were interested only in the isolated glands, which they now tried to approach, and not in the fluttering female moth, in which the loss of the glands is not at all noticeable to the human observer. Thus for these curious animals the female's attraction consists entirely in her scent.

Tasting with the toes

In insects the sense of taste is generally located in the mouth, and, as with us, it is less acute than the sense of smell. Again it does not need to respond to minute quantities of chemical substances over a distance, but serves to check up on the food while it is eaten.

There are, however, exceptions to this rule. Some flies and butterflies have very sensitive taste organs at the tip of their feet! If a fly running across the table steps into a drop of spilled jam, or if a butterfly, attracted by the scent of a ripe fruit, settles on a spot where the skin is broken, the sweet juice is noticed at once and the sucking tube unrolls.

We find something similar in certain fish. In the gurnard, a fish common along the Mediterranean coast, the foremost rays of the pectoral fins are converted into movable spines on which it crawls along (Fig. 37). The tips of these spines are covered with taste organs so that the fish at once realizes when something edible is at hand.

Man likes to consider himself the crown of creation. As far as taste and smell are concerned, animals certainly are his superiors.

10. *Feeling and Hearing*

Feeling seems to be a very different thing from hearing. Touching a doorknob and listening to the song of a blackbird — what could be more different? We would say they have nothing in common. And yet the design and development of the organ of hearing show us that it is nothing but an immensely delicate tactile receptor.

Remember the touch spots in the human skin. In hairy skin their position usually corresponds exactly with the roots of the hairs.

The hair acts as an auxiliary structure, increasing the efficiency of the receptor. Just as a worker can move a tree trunk with less effort if he pushes a lever under it, so the touch-sensitive nerve ending is effectively excited when the long lever, the freely protruding hair, is touched ever so lightly at its tip. Especially well-developed touch-sensitive hairs are the whiskers of cats. Through their length they are capable of supplying touch information from a distance. Thanks to them their owners do not bump their heads when they slip through bushes and undergrowth on their nightly forays.

Organs of equilibrium

By adding some further auxiliary structures Nature has made use of a similar principle in the design of the organs of equilibrium. They belong to the equipment of many animals, though not of all. They tell the animal, even in darkness or when suspended in water or air, what is up and down. How else could an eyeless jellyfish find its way about in the vastness of the ocean? We ourselves have these organs too, though usually we are not

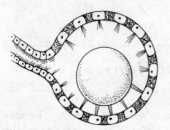

Fig. 38. Diagram of a balancing organ (organ of equilibrium) from the scallop.

aware of it. The sense of equilibrium does not figure among the proverbial five senses of Man.

In the simplest case an organ of equilibrium consists of a little fluid-filled sac, or cyst, provided with a layer of sensory cells. Fine hairs project from these cells into the fluid, and on them rests a solid body (Fig. 38). The force of gravity pulls this stone downward onto the cushion of sensory hairs. When the body tilts sideways the stone topples over and in doing so bends the sensory hairs. This signals to the animal the direction and degree of tilting, and may enable it to counteract the tilt by active righting movements. It is immaterial where the organs are located. Some animals have them in their tails and others in their heads. In the crayfish and its marine relative, the sea prawn, they lie in the feelers and are open pits that are filled, not with a crystal, but with grains of sand and other foreign particles.

About half a century ago, it was first discovered that these

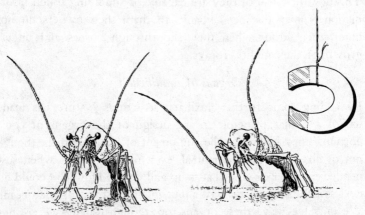

Fig. 39. Effect of magnet on the equilibrium of a crayfish. Iron filings have been placed in the cyst of the organ of equilibrium.

organs serve prawns in the maintenance of equilibrium and are not, as was thought till then, used for hearing. When the animal molts, the chitin in the pit is removed too and with it the grains of sand. Then the animal uses the fine pincers on its forelegs to push a new supply of sand into the pit. If the animal is put into an aquarium the floor of which is covered with iron filings, it picks up the filings instead of sand. If a magnet is now moved toward one side of the animal, the animal at once orientates its body according to that direction (Fig. 39). It behaves as if it had received compelling information that the whole aquarium was tilted onto one side.

Mrs. Cricket is called to the telephone

Yet another set of auxiliary structures combined with tactile receptor cells produces an organ of hearing. Such an organ is sensitive to sound waves, periodic vibrations that to us appear as sound or noise. They have their origin in a vibrating object and can travel through the air in all directions. When they are very powerful, such as a low note played loudly on the organ, you can feel them all over your body. Mostly they are too weak to stimulate the touch receptors of the skin, and only the organ of hearing responds. Its form and function are probably easiest to understand in the insects.

Quite a lot of insects have ears. They sit not on the head but on the abdomen and sometimes on the front legs (Fig. 40), as in crickets and the large green grasshoppers. They are not difficult to discover, provided one does not expect them to be as prominent as donkeys' ears. On the outside one sees just a little slit on either side of the leg (Fig. 40). It leads into a pocket that is lined by a thin, tightly stretched membrane; this is a kind of eardrum. Behind it lies an air space. The sound waves make the membrane vibrate, and the sensory cells nearby are stimulated by this vibration. What the animal really experiences is beyond our ken. That the organ perceives sound, however, and is there-fore an ear we know from the delightful experiments of a

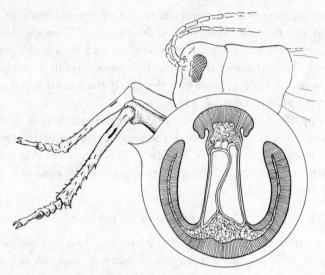

Fig. 40. A cross section through the organ of hearing of a grasshopper. The black spot on the leg locates the organ.

Viennese schoolteacher, Professor Regen.

Crickets, locusts, and cicadas are well known for their chirping concerts. In crickets and cicadas only the males chirp. That was already known to the ancient Greeks, and the somewhat malicious Xenarchos wrote: "Happily live the cicadas, for their womenfolk are dumb!" But the females are obviously not deaf, because they fly or run toward their chirping males, which may be said to serenade their sweethearts. But is the sound really the attraction?

In order to decide this Professor Regen put a chirping male cricket under a black cardboard cylinder, which had a small opening cut into its lower rim. The female followed those hidden sounds and after only a short search promptly entered the cylinder through the little opening.

But was it really the sounds that guided her and not, by chance, the scent of the male? Moths (compare Fig. 36) find their females by their scent. To decide this Regen let a captive male

chirp in one room and transmitted the sound through a micro-
phone into a distant room in which he kept a female cricket:
the female went to the telephone receiver — convincing proof
that it can only have been the sound that attracted her.

Communication by sound is quite common in insects:
beetles, butterflies, ants, and bugs alike. The musical instruments,
however, are very diverse. Some locusts draw their notched thigh
across the sharp edge of a wing; other insects rub their wings
together. Some bugs have a ribbed patch on their chest that looks
like an old-fashioned wash board, and if they want to be noticed
they work this strange harp with the tip of their sucking tube.
Thus, everyone makes music of a kind in his own fashion. Some
insects and many lower animals, however, appear to be deaf
and dumb.

We thread our way through a labyrinth

By ears we denote the more or less well-shaped appendages to
our head. They and the ducts leading from them into the head
are the entrances through which sound waves reach us. The
sense organ itself lies deeply embedded in the bones of the skull,
and its design is so complicated that it has been called a labyrinth.
Let us be brave and enter it!

There are two reasons why in Man and in all other vertebrates the inner ear is so complicated. First, it combines the
organ of equilibrium with that of
hearing, and, second, the equilibrium organ is not a simple sac, as in
the crayfish, but itself is subdivided into a number of parts of
different function (Fig. 41).

If we imagine the surrounding
bone removed from the labyrinth,
there remains the inner ear as a
thin-walled structure of a shape
that we see somewhat simplified in

Fig. 41. The organ of equilibrium (or labyrinth) of Man.

Fig. 42. Diagram of the semicircular canals.

Figure 41. It is divided into an upper and a lower unit. The upper sac with the three semicircular canals (Figs. 41, 42) is the organ of equilibrium; the lower with the spiral appendage serves for hearing.

On the floor of the upper sac is a group of sensory cells that are covered with calcareous crystals. This part of the labyrinth corresponds exactly to the organ of equilibrium in the crayfish and informs us about our position in space. With the help of the semicircular canals, however, we are aware of every turning movement in space, whether we ourselves turn or we are being turned, for example in a vehicle that takes a curve. How the stimulation of the sense organ comes about is not difficult to understand. When we rotate a basin with water, the water within it moves more slowly than the basin wall. The same happens when we put a flower pot upside down into that basin and thus create a circular water channel around it (Fig. 43). This also happens in a semicircular canal in the labyrinth each time we move our head. However, the semicircular canal has on one end a bottle-shaped enlargement into which projects a tender jelly dome containing delicate hairs that are linked to sensory cells below it. When we turn our head, the less rapidly moving fluid, similar to the water

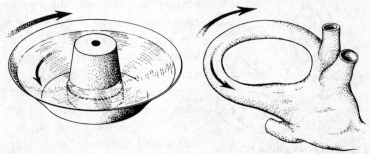

Fig. 43. How the impression of rotation originates.

in the bowl, pushes against the jelly dome and, by bending the sensory hairs in it, stimulates the organ. Thus we perceive rotations. In order to signal the direction of the turn it is important that each labyrinth have three canals in the three planes of space roughly at right angles to one another (Fig. 42). If we turn our head in a horizontal plane the fluid in the horizontal canals will move. It will not move in the vertical ones, just as the water in a basin does not move when we lift it up. During rotations in the vertical plane, however, the vertical canals are stimulated and the horizontal one rests.

Under normal circumstances our balancing sensations are not of crucial importance to us, since we have other means to ascertain our position and movements. We hardly pay any attention to them except when we overtax our semicircular canals by a whirling dance that makes us giddy. If, however, we indulge in underwater swimming, they do valuable service. People with a diseased labyrinth run the risk of drowning because they do not find their way back to the surface.

The auditory organ proper is the lower sac of the labyrinth and especially the coiled structure known as the cochlea. At the end of our auditory channel the sound waves meet the eardrum. Behind this lies the air-filled middle ear. As in the leg of a locust, the eardrum is a membrane stretched between two air-filled spaces which the sound waves cause to vibrate. But the sensory cells in this case do not lie close by: they are reached in a fairly elaborate way.

The vibrations of the eardrum are transmitted through a chain of three ear ossicles to the fluid that surrounds the labyrinth (Fig. 44). A window in the bone makes the connection, and a second window is closed by an elastic membrane. Fluids are not easily compressible. Every time the vibrating ear ossicles press inward, the fluid surrounding the labyrinth gives way at the only place where this is possible, by pressing the elastic wall of the other window outward and vice versa during a countermovement. The path from one window to the other leads

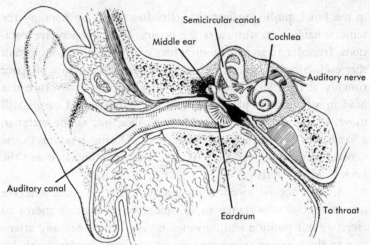

Fig. 44. Section through the auditory canal, showing ossicles and labyrinth.

through the cochlea, past its lower membrane, which starts vibrating and excites the auditory sensory cells, arranged all along it.

The higher the frequency of vibration, the higher the tone we hear. The bowed G string of a violin vibrates about two hundred times a second, the E string about six hundred and fifty times; accordingly, the number of sound impulses per second that stimulate our ear is different for each tone. We can perceive the difference of a very few vibrations per second as a distinctly different tone. There have been several attempts to explain this. One of the most fruitful theories was based on the fact that the basilar membrane contains a series of fine, transversely stretched fibers that get progressively shorter. They were supposed to respond by resonance to different tones in accordance with their length. Under otherwise equal conditions, a shorter string produces a higher sound. A violinist makes use of this when he shortens a vibrating string by pressing it down with the fingers of his left hand in order to change the pitch of the string. The tone produced by a vibrating string will make a resting string of the same length and tension vibrate. This can easily be verified by heartily singing a tone into an open unmuted piano. The string that is tuned to this tone starts to vibrate and the tone sung by the human voice finds resonance. Such resonance was held to be the basis for the tone discrimination by our ear. Only those sensory cells lying close to a certain group of fibers were thought to respond to a certain note. In recent years this theory has been modified, but it may still be assumed that different parts of the cochlea respond to different sound frequencies.

Of course, we can give no reason why vibrations of the air should give us the experience of sounds, and why, by their choice and manner of production, they can mean exquisite delight or grating dissonance. Fascinating theories have been developed about this, but the final truth is still unknown.

Can fish hear?

In most vertebrates the inner ear resembles that of Man. The labyrinth of fishes, however, has no cochlea, and it was therefore believed that they were deaf, and they really seem to be so. If we call to a fish it will not take any notice of it. In the normal life of a fish a human being is of no importance, so what reason should it have to listen to us?

However, some fish in fact hear very well indeed. I once had a blind catfish. He got very popular with my collaborators and friends. Usually he lived in a hide-out on the bottom of his aquarium. Then I had an idea that was unexpectedly successful. I wanted the catfish to come out of his hide-out in response to whistling. It took only a few days before he had grasped that whistling meant food. Every time I whistled he came at once to the surface in search for food.

By this method of training, known as conditioning, the power of hearing could be proved and studied in detail in many fish. It was found that there are fish that are hard of hearing and some that hear very acutely, such as catfish, carplike fish, and others. They can hear sounds so soft that we can hardly perceive them. One might ask why this is so. Fish, it is generally thought, are dumb and have nothing to tell each other. But one knows many kinds of fish that produce grunting, low-sounding, squeaking, grating, and whistling sounds, and much too little attention has been given to their means and manner of sound production. Here a wide field opens up for the study of animal communication.

Fish are, within limits, capable of distinguishing different tones. One can prove this by first training them to respond to a tone of a certain frequency blown on a whistle. As soon as the fish has learned that this sound means food one blows alternately with the conditioning tone another one, maybe a higher one, at which no food is offered. At first the animal will expect food at this sound too and will start searching for it; in this case it is warded off by a tap with a glass rod.

After a while it appreciates the difference and searches for food in response to the food tone, while it goes into hiding when the warning tone is sounded. By slowly diminishing the difference in frequency between the two tones one finds out which interval a fish can still distinguish with certainty. Usually it is about an octave. If one offers the two tones in rapid succession instead of after a certain pause in time, then they can distinguish with certainty even a difference of a quarter tone. This is not a very striking achievement as compared with the excellent sound discrimination of Man or dog. After all, the ear of fish still lacks a tuned cochlea.

It has been experimentally proved that in fish too sound perception is a function of the labyrinth, in fact of the lower sac from which the cochlea has evolved in higher vertebrates. However, instead of a basilar membrane, it contains only two calcareous stones that are in contact with groups of sensory cells that are stimulated by sound waves. In fish with an acute sense of hearing a complicated auxiliary structure ensures that the vibrations reach the calcareous stones. There is, however, no specially tuned apparatus that helps to distinguish frequencies. All sound frequencies meet the same sensory cushion. This compares to the sensation we get when we put our finger tips lightly on a vibrating object. In this way we too can distinguish vibration frequencies within limits. The ear of a fish essentially perceives nothing more than do the touch receptors in our finger tips. It is something intermediate between a touch receptor and our ear. This again shows that feeling and hearing are very closely related indeed. Experience with deaf and dumb people leads to the same conclusions. They can learn to substitute their sense of touch for their missing sense of hearing and to develop its acuteness by careful attention and exercise. Within limits they can feel the sounds of the speaking voice by touching the throat or chest of a speaker with their fingers. The famous Helen Keller, who was both deaf and blind, had the piano or organ played to her and, on her own authority, seems to have enjoyed the sensation of musical vibrations. Of course she had no means of comparing her sensations with the true hearing of a piece of music.

11. *The Eye*

Nobody doubts that we see with our eyes. It seems just as obvious that creatures without eyes must be blind. But do not let us draw hasty conclusions. Have you ever looked into the eyes of an earthworm? You can search its head and the whole length of its body, but you will not find any. And yet this light-shunning creature can distinguish very well between light and darkness. If light falls on earthworms they quickly creep away and stop only when they reach a dark place. Under the microscope one discovers in their skin single sensory cells that look exactly like the light-sensitive cells in the eyes of other worms. These are the simplest light-sensitive organs we know, and one can hardly call them eyes.

From the eye spot to the lens eye

A simple device that improves the efficiency considerably is the one-sided screening of the sensory cells by a black pigment. Such eyes occur in little worms that populate our rivulets and ponds and that usually hide away under stones. Fig. 45 shows their black eye spots. The black pigment cup is the auxiliary device; the light-sensitive cells lie inside the cup. Light that comes from in front stimulates them. Light from behind is barred by the pigment screen. This is an advance over the unscreened cell because it indicates the direction of light, which in turn helps the worm in seeking a dark hide-out.

By slightly different means a similar effect is brought about in marine snails. The sensory cells lie in larger numbers in a sunken pit in the skin of the head and are screened off at the sides and in back by a black covering. In Haliotis, the shell of which is often used as an ash tray, the pit is fairly deep and its opening very narrow (Fig. 46). This seemingly unimportant variation may originally have been meant to give the sensory cells greater protection; actually it opens up quite new possibilities for perfecting the eyesight. In another mollusk, the Nautilus, the opening into the eye is reduced to a small hole. Nautilus is the last survivor of the ammonites, which, as shown by fossil

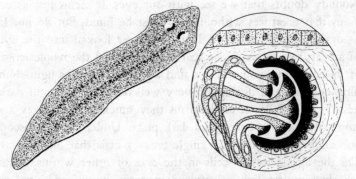

Fig. 45. A planarian, or flatworm, with eye spots. Cross section through an eye spot. (Greatly enlarged)

finds, had a very wide distribution in earlier periods of the history of the earth. It was Nature that first invented the pinhole camera, long before the human mind thought of it some mere four hundred years ago. The pinhole camera is a black lightproof box with a small opening in front. As light rays travel in a straight line, each object outside will produce a reversed picture inside at the back of the box. If one replaces the back of the camera by a ground-glass screen one can see the image. If one replaces the ground glass by a light-sensitive photographic plate the image can be photographed. The eye of Nautilus is such a pinhole camera: instead of a photographic plate there is the retina of the eye, in other words countless light-sensitive cells that are responsible for the reception of images.

This type of eye not only perceives light and darkness, or the direction of the light, but it produces an image of the surroundings.

Pinhole cameras have gone out of use for a simple reason: if the hole is made very small, the result is a sharply focused but dim picture. If we widen the opening the picture will be brighter but blurred, because from each point of the object a whole bundle of light rays will now be able to enter, which will no longer produce just a point of light but a whole patch. Adjoining light patches then will overlap and the image become blurred.

In the lens camera these difficulties are overcome by widen-

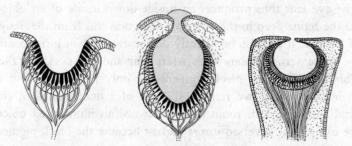

Fig. 46. Forerunners of the lens eye. Cross sections left to right of eye of Patella, of Haliotis, and of Nautilus.

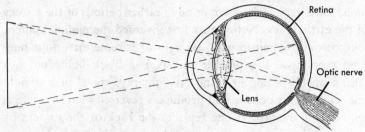

Fig. 47. Cross section of the human eye.

ing the aperture and fixing a lens in it. Light rays emitted from one point are refracted in such a way by the lens that they are reunited on the screen into one point. Now the picture is both clear and bright. Nature made this invention too, long before the advent of Man on earth. The opening of the eye got widened and a lens was put behind the pupil. We find lens eyes in many worms and snails, and among all the vertebrates including Man (Fig. 47). The pinhole camera with its narrow opening, which works efficiently only in very bright light and is ineffective in dim light, has been handed down to us through Nautilus as a reminder that no great task succeeds at the first attempt and that gradual development toward perfection is not only the hallmark of the scheming human mind but a general characteristic of living Nature.

The world topsy-turvy

The eye lens thus produces an upside-down image of an object on the retina deep in the eye. We can guess this from the design of the eye. It can also be directly demonstrated. Not infrequently we come across rabbits with white hair and red eyes, so-called albinos, which lack black pigment not only in their fur but in their eyes too. If we remove the eye of a dead animal of this kind and in a dark room hold it toward an illuminated object we can see the reversed image clearly because the black pigment screen is missing.

Our own eye too produces a reversed image. One often

hears the question: Why then do we not see everything standing on its head? But this is nonsense. It is not in the retina of the eye but in the brain where we become conscious of the image. There the parts of the picture are quite differently arranged according to the pathway of the nerve fibers that connect the light-sensitive cells with the nerve cells of the brain (Fig. 47). From our earliest youth we depend on the concerted help of our visual impressions and our touch sensations for our appreciation of observed objects in their relative position to each other and for our experience of three-dimensional space as such. We know from the aimless movements of very young children and their unsuccessful attempts at touching objects that they have as yet no correct idea of space. At the same time we observe their quick progress through daily experience. It is a pity that none of us can remember how we saw the world during the first days of our life, and philosophers try in vain to discover how the excitation of a nerve cell in the brain can lead to our conscious experience of space. The human mind can never understand and explain itself.

Things as we see them are not simple copies of the images projected onto our retinas but the result of a mental process. We receive two retinal images, one from each eye, but they are not identical because we see objects with each eye from a slightly different angle. We can easily convince ourselves of this by holding our head quite still and looking at a near object first with one and then with the other eye. The two different images form in our mind one single stereoscopic image. That we see objects in three dimensions, that we judge their distances correctly and are capable of depth perception, is mainly due to the difference in the two retinal images. This difference is proportional to the distance in space. This much we know, but we do not understand the processes by which our mind interprets it.

Accommodation; far- and nearsightedness

The sensory cells are the intermediaries between the image formed by the lens and our perception of it. They are small cells,

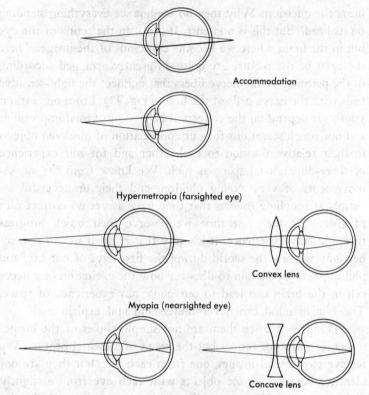

Fig. 48. Kinds of accommodation of the human eye. Above, for the farsighted eye; below, for the nearsighted eye.

each one so slender that several hundred of them are packed, in an orderly fashion, along 1/25 of an inch. The depth of this layer is of microscopic dimensions too. The image projected by the lens normally falls within this sensory layer; if not, it appears blurred (Fig. 48). To understand this you have only to imagine a photographic camera in which the lens projects the image exactly onto the light-sensitive layer of the film. If a distant object is right in focus it will get out of focus when it is brought nearer to us. The light rays that come from a nearer source meet behind the photographic plate producing on the plate itself a blurred patch of light. The image is out of focus. The photographer overcomes this difficulty by changing the position of the lens. He draws it out in order to lengthen the distance between the lens and the photographic plate until the image of the near object is again in focus. The corresponding process in our eye is called accommodation (Fig. 48).

There are in fact eyes that work in this manner, the distance between lens and retina being changed when a near object is to be brought into focus. Frogs do it that way. A muscle in their eye draws the lens forward. The same can be achieved if the refractory power of the lens is changed by an alteration of its shape. This is the case in the human eye. When we look at a near object the curvature of the elastic lens is increased by the action of certain muscles. If, with advancing age, the elasticity decreases the mechanism of accommodation breaks down, and near objects can no longer be sharply focused on the retina. An old man trying to read the paper formulated this predicament very aptly by saying: "My eyes are still all right, it is my arms that are too short." We cannot make our arms grow but we can wear spectacles which, by making the light rays converge, make up for the refractory power lost by the lens.

Farsightedness can occur even in young people. The developing eye may fall short of its normal length. The distance between the lens and the retina is then too small and the image of distant objects will come to lie behind the retina. Active accommodation will, therefore, be necessary even for distant objects, and this makes focusing on objects close by still more difficult. On the other hand an eye may develop so as to be too long, and its owner will be nearsighted. The image of a distant object will be blurred because it comes to lie in front of the retina (Fig. 48). Only with the object so close that the normal eye would have to accommodate will its image become sharp. Nearsightedness can be compensated by spectacles. They have to be ground differently so that the light rays are made to diverge instead of converge.

The realm of colors

What would a sunset be like without its display of colors! That
we see the world in colors has not been mentioned so far. Just
as we perceive sound waves of different wave lengths as tones of
varying pitch, so we perceive light waves of different wave
lengths as different colors. Sound waves and light waves are, however, physically different. The former are mechanical disturbances of the air or other tangible media; the latter are electromagnetic disturbances requiring no medium. What both have in
common is that the sensory impressions created by them vary
with wave length.

A mixture of light rays of different wave lengths, such as
the sunlight, is perceived as white light. If we send white light
through a prism, the rays are sorted out according to their wave
length and form a colored spectrum. The longer wave lengths
appear as red light, the shorter as violet (Fig. 49). In between lie
the other colors of the rainbow. The rainbow itself is formed in
a similar fashion. Here the sunlight is refracted by raindrops. The
surface colors of most objects, however, are brought about not by
refraction but by the fact that some of the wave lengths of white
light that falls on them are absorbed, others are reflected. Thus a
flower appears red to us if it reflects predominantly the long
wave lengths of daylight but absorbs the short ones.

The color vision of bees and the color of flowers

The colors of flowers are among the most beautiful and bril-

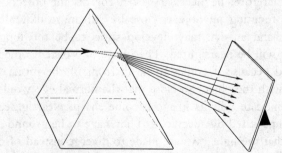

Fig. 49. Refraction of rays of light by a prism.

liant in Nature. It is, however, striking that not all flowering plants produce blossoms, but only those that produce nectar and are visited by bees and other insects. Insects collect the sweet nectar not as robbers but as welcome guests. They render an important service to the flowers although they are unaware of it. By carrying pollen on their bodies from flower to flower they bring about fertilization, without which seeds would not form.

In grasses, coniferous trees, poplars, and other plants pollen is transported haphazardly by the wind, and only the production of a very large quantity of pollen safeguards fertilization. These "wind-loving" plants have very insignificant flowers that produce no nectar. "Insect-loving" flowers have strikingly gay colors and are scented. It is reasonable to assume that the colored petals of flowers are an advertisement, showing bees and other winged creatures from afar where something good is to be had and where their visit is required.

Can bees see colors as we do?

This can in no way be taken for granted. The insect eye is completely different from the human one. On each side of the head of a bee or fly are several thousand tiny eyes (Fig. 50), each of which perceives not the complete image of the environment but only a small portion of it. Since there are so many of them, each looking in a slightly different direction, the many simple units give the effect of one compound eye with the many individual portions fitting together, mosaiclike, to give a complete image of the surroundings. Whether they can perceive color one cannot conclude from their design; this one has to find out by experiment.

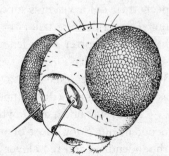

If we feed bees with sugar water set on a blue piece of paper they collect as eagerly as if it

Fig. 50. Compound eye of the fly composed of many tiny eyes.

were nectar out of flowers. After they have come and gone several times we move the food and put down a clean piece of blue paper and also a red one. The bees settle down only on the blue. They learn that blue means food and show that they distinguish red from blue. However, this does not yet prove color vision. There exists a rare eye disease in human beings known as total color blindness. To people afflicted with it a colorful landscape looks like a black and white photograph. A certain light intensity corresponds to each color. Therefore, such people too can distinguish red from blue because red appears to them fairly dark and blue relatively light. In order to exclude the possibility of bees recognizing colors just by their light intensity, we have to make a change in our experiment.

Again we feed over blue; but now we surround the blue paper with a variety of gray papers whose light intensities range from white to the deepest black. Under such conditions a totally color-blind eye cannot spot the colored paper with certainty because it appears as a gray of a certain density very similar to one or the other of the gray papers around it. Bees, however, still choose their blue from among the many shades of gray without hesitation. This is how they prove to have true color vision.

Their color vision is not quite the same as ours. If we train them to respond to scarlet red they will seek food not only on the red but also very eagerly on black and dark gray papers. To bees red and black appear the same; they are red-blind. In other aspects their color vision is superior to ours. Bees not only perceive but see as a definite color the ultraviolet rays in sunlight, which have a still shorter wave length than the violet ones. We cannot see them and have had to wait for physicists to discover them. What ultraviolet looks like to them exactly we can imagine as little as the other things that happen in the mind of bees.

Now we understand why scarlet red flowers hardly ever occur in our flora. Most of our red flowers like heathers or red rhododendron and red clover are bluish red and have been shown to appear to bees as blue, as they cannot see the red element.

Poppy flowers are scarlet red. The poppy flowers, however, reflect a lot of ultraviolet light, and the eye of the bee sees the petals in shining ultraviolet. This has been proved by experiments. Scarlet red flowers occur in many ornamental plants introduced from abroad. In the tropics, where they come from, they are not pollinated by bees but by birds like the hummingbird and honey bird, which, hovering in front of the flowers, collect nectar from them. According to recent investigations red is to the bird's eye a color of a specially high brilliancy.

This shows that the colors of flowers did not originally evolve to please Man but to catch the eye of their animal visitors. We enjoy their beauty none the less.

The perception of polarized light and the sun as compass

It is only a few years since we have known that the eye of the bee differs from ours in yet another way: it perceives the direction of plane-polarized light.

Again, we know of polarized light only through the efforts of the physicists, not from sensory experience. We learned that light waves oscillate in a direction transverse to the direction of their propagation and that in natural sunlight the plane of oscillation changes rapidly and at random, while in plane-polarized light the oscillations are confined to one plane only (Fig. 51).

Polarized light enters our eyes on many occasions. When sunlight is reflected from a wet road the plane of polarization lies parallel to the road surface. Polarized light comes from the blue sky; the direction of the plane of polarization varies, however, in different regions of the sky. And because it is influenced by the position of the sun, it also varies specifically in one and the same place according to the hour of the day. Since our eyes are

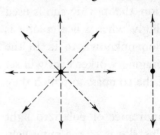

Fig. 51. Polarization of light. Left, looking at an unpolarized beam; right, at a plane-polarized beam.

not equipped to register the orientation of plane-polarized light, we can observe it only with the help of certain filters. Bees and other insects, also spiders and crustaceans, observe it directly. Their eyes have built into them microscopically small devices by means of which they can discern the direction of polarized light.

This is very important for orientation. When we want to follow a certain direction we use a compass. Animals have no such instrument, but they use the sun instead. The sun is very reliable except that it changes its position in the sky throughout the day. Therefore it is only useful for orientation purposes if one can tell the time of the day.

Experiments with bees have shown that they have an unbelievably accurate sense of time. When one feeds a few specially marked bees at an experimental feeding place at a certain hour of the day and continues this for several days, the bees appear at exactly the right time at the feeding place, even if no food whatever is put out for them. The same happens whether the experiment is carried out in a room or deep down in a mine. It is, therefore, not the sun that helps them to read the time but a clock built into their own body. What this clock is we can only guess.

Thanks to the inner clock, the bee not only knows the time but knows, however surprising this may sound, the appropriate position of the sun for every hour of the day. When trees or other landmarks cannot be used for orientation, the moving sun is used as a compass. This appears most strikingly when a bee colony is taken over long distances into a region unknown to it. In the completely unfamiliar surroundings foraging worker bees will set off to look for food in the direction of the compass to which they are accustomed.

When the sky is overcast the importance of polarized light becomes obvious. A small patch of blue sky is of the same help to the bee as the sun itself because it reveals the plane of polarization and with it the position of the sun.

Bees are a very highly developed group of insects, but this is not relevant here: the perception of polarized light occurs widely among the arthropods. A little insignificant crustacean, the sand hopper, which lives in the tidal zone of the seashore, orientates itself exactly like the bee, and when a strong wind blows it onto the dry sand it finds its way back to the sea with the sun as compass and the polarized light of the sky as guide.

12. *How Nerves and Hormones Co-ordinate*

It is the task of the nervous system to co-ordinate the activities of the multitude of cells and of the organs in the body. The highest center of co-ordination is the brain. In the brain all the messages from the sense organs are received and sorted out, and from there the directives for action are sent to the effector organs. The nerve cells, an astronomical number of which make up the brain, thus form the connecting link between all parts of the body. The nerve strands are the pathways along which the messages travel, and they have been compared with telephone cables. Our spinal cord is, as it were, a subordinate center to which the brain delegates certain tasks that can be done independently and without interference from above.

Reflexes

Subordinate executives usually have to follow rigidly fixed rules of action. The same holds in the nervous system. The strictly prescribed actions of the nervous system are reflexes. It is characteristic that a certain external stimulus releases a predictable reaction and if the same stimulus is applied twenty times it will elicit twenty identical responses.

If one lifts up a beheaded frog it does not move at all, its legs hang down limply and it looks dead. Cold-blooded animals,

Fig. 52. Diagram of a reflex arc in the frog. The stimulus consists in pinching the foot.

however, do not die instantaneously. As the beheaded frog has no brain it is reasonable to assume that it has lost awareness. We find, however, that its organs, including the spinal cord, remain alive for hours. If we pinch a foot the leg is pulled upward in a short but powerful movement. By the pinch the sensory cells in the skin were stimulated, the stimulus traveled through the nerve fibers of the leg to certain nerve cells in the spinal cord, and from there through other nerves to the muscles of the leg, thus bringing about their contraction (Fig. 52). This is a reflex, which can be performed repeatedly as long as the pathway along which it travels is in order. Without the stimulus a decapitated frog would never draw up its leg.

But why did we actually decapitate the frog? Without this drastic measure we could not have studied the reflex properly, because the reflex arc itself is linked through special nerves with the highest center, the brain, and can be influenced by it. If we treated the normal animal as we did the torso, it would continuously pull up its legs and wriggle and behave as if it grasped the situation. Sometimes a reflex action can be inhibited by the brain because of other stimuli. We have to exclude the primary center when we want to demonstrate a pure reflex action.

Reflexes play an important role in the life of lower animals, and they occur in higher animals and human beings too. We all know about the movements of the pupils of our eyes. In dim light the pupils are wide open and admit a maximum of light. If we enter a brightly lit room or shine a torch into our eyes the ring muscles in the iris contract, reduce the pupillary opening, and thus protect the eye against too strong a light. This happens so rapidly that one might assume the light made the iris respond directly. And yet we are dealing with a reflex that is routed through a subordinate part of the brain — the brain too has higher and lower centers — and that fails to occur in blind people or when the reflex arc is interrupted at any point.

We are not aware of our pupil changing its diameter except when we observe it in the mirror, and the process can in no way be influenced by will. Many such reflexes are known. There are others of which we are conscious and which we can therefore suppress by will power. For instance, we can keep ourselves from blinking when an object is moved quickly past our eyes.

Instinctive activities

Inborn urges and reflexes can combine and give rise to complicated actions, which we call instinctive. They look like purposeful activities but are in reality just as much a part of a hereditary pattern of behavior and independent of individual experience as the simple reflexes. Instinctive behavior is innate, and we find it most highly developed in insects. Their activities appear often so purposeful and sensible that the naïve observer may be induced to admire their intelligence.

The yucca plant, a native of America, develops its yellowish white flowers at a time when the little yucca moth is mature. The female moths fly to an open calix and collect a large ball of pollen, which they carry away, pressing it against their neck. Thus laden, they fly to another flower and stuff the pollen into a dent of the receptive stigma. The growing pollen tubes fertilize the ovules of the plant and seeds set. A botanist wanting to grow the yucca seed could not do any better. The moth even behaves as if it knew that better seeds result from cross-fertilization than from putting the pollen on the stigma of the same flower. Is the development of seeds of any importance to the moth? Yes, it is absolutely essential not for the moth itself but for its brood. It lays its eggs into the very flower in which it has brought about cross-fertilization. The hatching larvae live on some of the developing seeds, while the remainder of the seeds are left to guarantee the survival of the yucca plants as a source of food for future generations of moths. The grown-up caterpillars pupate in the soil and hatch only when it is time for the yucca plant to flower again. We frequently grow yucca as an ornamental plant. But in our gardens the moth is absent and we never get any seeds. Thus the plant is entirely dependent on the moth. The complex behavior of the moth looks in every detail purposeful, designed, so to speak, to ensure a food supply for the offspring. Yet it does not act from insight. When it approaches a flower it is naturally ignorant of the meaning of fertilization and it never lives to see the result of its activities. Its behavior is innate, or instinctive.

The candidate that failed the intelligence test

That instinctive actions of insects are not at all coupled with insight can be most spectacularly shown when the animal is faced with an unaccustomed task. This can be illustrated by the following example. The leaf-cutting bee is a near relative of the honeybee but lives a solitary life. For each egg, the female builds into old tree trunks or beams a thimble-shaped nest made of bits of leaves. The bee uses its jaws to cut out two different patterns from the leaves: oval pieces for making the thimble and circular pieces for the lid and for the stopper with which it closes the whole nest. The circular pieces of leaf are suitable only for the closure of the usual cylindrical passage into the nest. It has been recorded that in a case where the entrance into the nest, instead of being cylindrical, was formed by a gaping cleft, the bee did not deviate from the circular pattern of leaf cutting. Instead, it tried to plug up the cleft with circular pieces which it had to drag into it vertically side by side, in a longitudinal instead of a transverse arrangement. Thus a lot of space was left open to possible intruders ón either side of the inefficient plug. Instinctive action serves well when things happen according to rule. Exceptional circumstances are bound to produce nonsensical solutions.

The stickleback as bridegroom

Instinctive actions occur also in vertebrates, and they are much more common than one usually assumes. A special study has been made of the stickleback, a small fish prevalent in certain waters and interesting for its strange habits. In the mating season the male makes a nest from water plants and changes into resplendent wedding attire: the belly and the sides are set off in a luminous red against the blue-green shading of the back. By instinct the fish attacks fiercely every rival that dares to approach its territory. How does it recognize the other male and how does it distinguish it from other indifferent passers-by? We can check on this by confronting the lovelorn stickleback with artificial models. It was found that no notice is taken of the most accurate copy of a male if it lacks the red belly. A model that hardly looks like a fish is, however, attacked if it has a red underside. The picture of the rival is thus simplified and restricted to the one special characteristic. The red belly triggers the instinctive fighting behavior.

For the recognition of the female too the fine contours of its body are inessential. It is its belly swollen with eggs that releases courtship behavior in the male. What follows is a ritual of strictly prescribed patterns of alternating activities by each partner, and these are triggered one by one in certain succession. The end result is the laying and fertilizing of the eggs. It makes sense that simplification of triggering stimuli for complex instinctive activities not only eases but makes possible their hereditary transmission.

Instinctive actions are part of human behavior also. The newborn child seeks and finds the breast of its mother and does not need to learn how to use it. Even in grownups instinctive behavior persists, though reflection and intelligence gain more and more the upper hand. Complex instinctive actions may have evolved from a combination of reflexes; it is, however, impossible to find a bridge leading from instinctive to intelligent action in human beings and higher vertebrates. The latter may have evolved from reflexes but along a different line.

Conditioned reflexes

If we made the mistake of drinking a caustic fluid our gullet would be badly damaged and we would find it impossible to swallow. The surgeon would rescue us by making a little opening into the wall of our abdomen and establishing through it a direct connection with the stomach. Through this artificial feeding passage food and drink can be taken in. Such an opening is called a gastric fistula.

The Russian physiologist Pavlov carried out this same operation on dogs in order to study the secretion of gastric juice. So as to get the gastric juice pure, without any food particles mixed up with it, he also cut the gullet and sewed the two openings to the skin of the neck. As soon as the wounds had healed the dogs ate with great appetite and did not notice that everything fell out again through this opening. As soon as they started feeding, the stomach produced large quantities of digestive juice, although no food reached it. The secretory activity of the stomach is a typical reflex. The meal stimulated nose and palate, and the gland cells of the stomach lining were activated via the brain centers. At the end of the experiment the dog was, of course, fed through the gastric fistula.

From this nothing new might have been learned apart from the fact that nerves regulate not only the activity of the muscles but also that of glands. But Pavlov made a strange discovery. If, in the presence of the dog, a key was struck on the piano there was no reaction. But if the key was struck on the piano whenever the dog was being fed, after a while the gastric juice started to flow at the striking of the key alone. This means that after repeated simultaneous excitation of nervous pathways connected with feeding and hearing a connection becomes established in the nervous system which links perception of sounds with gastric secretion. Pavlov called this the establishment of a conditioned reflex because its appearance depends on certain conditions, in this case on the repeated and simultaneous offer of food and sound. If these conditions do not prevail for a prolonged period, then this acquired reflex becomes extinct again.

A conditioned reflex is not a great mental achievement and it is by no means restricted to the animal mind. It also happens in Man. To mention one example: a child with a gastric fistula was played a tune on a trumpet at mealtimes. After a short time the tune alone produced a secretion of gastric juice, just as in Pavlov's dogs.

At the root of intelligence and insight

Since the first discovery of the conditioned reflex a great variety have become known. Instead of sound any other sensory perception can be associated with it, and instead of glands, reflexes of muscles can lead to conditioned movements. They are known in all lower animals, including insects, but especially in vertebrates.

The conditioned reflex may probably be considered as the bridge between innate reflexes and simple intelligent action. If a brain becomes capable of rapidly establishing conditioned reflexes and, at the same time, of processing simultaneous sensory impressions and drawing on the storehouse of memory, it arrives at an interplay of thought that seeks and finds relations and puts experience to profit. This is the basis for rational behavior. In the highest vertebrates this is developed to a fantastic degree and is not confined to Man alone.

The psychologist Wolfgang Köhler, who experimented with chimpanzees on the island of Tenerife, observed their rational behavior on many occasions. To give one example: a chimpanzee sits behind the bars of its cage; out of reach in front of the cage lies a banana. The animal takes a stick and uses it to hook in the fruit. Another time it has no such stick. Now it goes to a tree inside the cage, breaks off a branch, and uses it in a similar way.

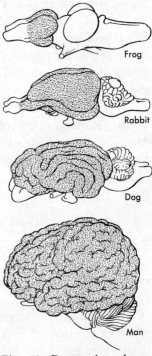

Yet another time the ape has at his disposal two hollow bamboo sticks, each by itself too short to reach the coveted fruit. It does not occur to the animal to fit the sticks together. After several attempts with the short sticks it gives up. A little later, as the animal is playing with the sticks, it fits them together by pure chance so that now it has a long stick. Suddenly it has a "flash of insight": it runs to the bars and uses the doubled stick to hook the banana successfully. From that moment onward, confident and without hesitation, it proceeds immediately to fit short-length sticks together. An invention has been made and is purposefully applied.

Any dog lover will be able to relate similar intelligent exploits of his pet, different in degree but not in kind.

Fig. 53. Progressive change (evolution) of the forebrain. The cortex is shaded.

The more highly complex mental activities of the higher vertebrates find their anatomical expression in the enormous development of the cerebral hemisphere of the forebrain (Fig. 53). Research reveals a perplexing multitude of nerve cells and nerve fibers in the forebrain, but we cannot as yet imagine the relationship between the structure of this brain and the workings of the mind in animals and Man. We can see that the conditioned reflex leads from the rigidly inherited reflex to freedom of action. But it is a long way from there to the freedom of the human will and to the contents of our rich consciousness.

Hormones

The co-ordinated function of the body parts is not brought about by the nervous system only, but also by the action of chemical substances that are produced by certain glands and are secreted into the blood stream, which conveys them to the various effector organs. Because these substances activate the function of organs they are called activators or *hormones* (from the Greek "I activate"). The glands are called endocrine glands and are distinguished from other glands by the absence of glandular ducts. Their secretion is not discharged to the outside, like that of the sweat glands or milk glands, but is poured into the blood.

The females of mammals feed their newborn on the secretion of their milk glands. This secretion starts as soon as the young are born. One might believe that, similar to the secretion of gastric juice, this process is regulated by the nervous system. But the secretion begins immediately even if all nerves in the breast are destroyed. In guinea pigs milk glands were removed and transplanted into the ears, where they healed in so successfully that when the young were born the transplanted glands started secreting milk. This is a striking example of the co-ordination of vital processes through hormonal control.

In this case the hormones are produced mainly by a structure in the ovary that undergoes a change at the time of birth. They are distributed by the blood and in due course reach the transplanted milk glands and stimulate them into secreting.

The discovery that diabetes is due to the lack of a hormone was of vital importance. In this disease the surplus carbohydrates are not stored in the liver but stay in the blood in the form of surplus sugar, which is finally voided in the urine. The continuous loss of valuable fuel combined with other metabolic upsets can be fatal. The first insight into the true nature of the disease was due to a chance observation obtained in an experiment with a different end in view. Minkowski and Von Mering removed the pancreas of dogs in order to study its role in digestion. The animal keeper noticed that the urine of the operated animals attracted swarms of flies; it contained sugar. Unintentionally, the investigators had, for the first time, produced diabetes artificially. However, the gland cells that produce the pancreatic juice have nothing to do with this. Between them lie groups of gland cells embedded in the pancreas like islands in a sea which for a long time had been completely overlooked. They liberate a hormone into the blood stream to which the name "insulin" has been given. If secreted in adequate quantities, insulin regulates the sugar metabolism. If, however, the islets are removed with the pancreas or if they degenerate through illness, the hormone is absent and the body suffers from the effects of high blood sugar. In 1921 the Canadian scientists Banting and Best succeeded in producing insulin from pancreatic extracts of animals the hormone of which is useful in helping human sufferers from diabetes when injected into their blood stream. Although the destroyed islets cannot be restored, the necessary daily dose of insulin can be supplied to stop the disease. It is fortunate that hormones are not specific to a certain organism. The treatment is successful whether the insulin used comes from fish, cattle, or Man.

While the islet cells are hidden away within the glandular tissue of the pancreas, other endocrine glands are conspicuous organs and have been known for a long time. The adrenal bodies, for instance, lie in the neighborhood of the kidneys and produce adrenalin. When liberated into the blood stream, this hormone raises the blood pressure, stimulates the activity of the heart, leads to an expansion of the air passages in the lungs, restricts the activity of the digestive organs, and transports stored sugar from the liver into blood and muscles. These are manifold effects all working toward one aim: to raise the energy output of the body during strenuous effort. Adrenalin is a chemically well-known substance and can now be synthesized. It is effective in incredibly small amounts. If one wanted to dilute one gram of adrenalin to such an extent that an injection into the blood would cause no reaction one would need a train one and a half miles long, consisting of three hundred tank cars to carry the necessary water.

Not all hormones work as fast as adrenalin. Others influence the metabolism much more slowly but no less effectively. They regulate the growth and development of organs and of the body as a whole. Among them is the thyroid, the only endocrine gland that can on occasion be noticed from outside. We know an enlarged thyroid as a goiter. The predisposition to goiter is bound up with certain geographical regions and the reason is probably as follows: the thyroid hormone, called thyroxine, contains chemically bound iodine. Iodine is found in small quantities everywhere on earth and we take in traces of it with our drinking water. Regions in which goiter is endemic are lacking in iodine. This lack causes the formation of thyroid tissue of low quality which the body tries to compensate for by producing an enlargement of the gland. If infinitesimal quantities of iodine are regularly introduced in the diet, the natural lack of iodine can be corrected. Cooking salt that has been enriched with traces of iodine has been used with good results. But what is the function of the normal thyroid gland? This became dramatically apparent when too much of the thyroid was cut out in the surgical removal of goiters for the relief of breathing difficulties. Especially in young individuals the consequences were appalling. They remained small in stature and restricted in mental development to the point of cretinism, conditions that are unquestionably the effect of a lack of thyroxine. If a piece of living thyroid gland is implanted into any part of the body, or doses of thyroid are injected or taken orally, normal development is resumed and the bodily and mental deficiencies disappear.

These are only a few examples, meant to introduce us to the nature and function of hormones and to the endocrine glands which secrete them. There exist a great variety of hormones, with complex and far-reaching effects. The giants and dwarfs that we see at a freak show may not even know that their abnormal size is linked to the size of a small gland at the base of their brain. Certain kinds of mental deficiency and lethargy as well as violent passion may have their origin in endocrine glands, and as animal experiments show, they can be steered like a car in one direction or another, accelerated or braked. The courts have to pronounce judgment on human frailties and shortcomings, as a social necessity. They should, however, remain aware to what degree our actions are influenced by the physical constitution of our body, for which we cannot be held responsible.

Hormones in plants

It is only quite recently that the important role that hormones play in the life of plants has been discovered. The best-known hormone is auxin, which controls the growth of plants and can now be produced in a chemically pure form. Its effect appears most strikingly in plant movements that are due to unequal growth and that make the plant bend in one direction or another.

Growing plants turn toward the light. If light is directed on a seedling from the right side, it will start bending toward the right within two hours, because the part just behind the tip will grow faster on the shadow side (a in Fig. 54). The curvature does not appear if one cuts the tip off or if one covers it with tinfoil (b). The stimulus, therefore, acts at the tip and its effect is conducted toward the inner parts, which respond by curving. This conduction is not nervous but due to diffusion. This has been proved to be so by means of very elegant experiments. If the tip is cut off and then put back, conduction occurs across the cut (c) even

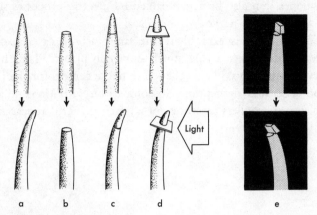

Fig. 54. Demonstration of the effect of growth hormone on the oat seedling.

through an intervening layer of gelatin (d). The substance that diffuses is the hormone itself. It is continuously produced in the cells at the tip of the growing plant, migrates downward, and stimulates the cells deeper down into growth. One-sided stimulation shifts the hormone, called an auxin, toward the side that is in the shade, and this consequently shows the intensified growth. In the experiment, hormone diffuses in fact through the gelatin plate and can be collected in it. If one then covers a capped, nonilluminated seedling with such a piece of gelatin, the hormone can penetrate into the seedling only at the spot where the gelatin plate covers it, with the effect that only this half of the seedling begins to grow and bends toward the uncovered side (e in Fig. 54).

Thus plants prove, broadly speaking, to be sensitive organisms too. They can turn toward the light and can respond to other stimuli also. So, under the influence of gravity, root tips grow downward. Insectivorous plants perform prehensile and enfolding movements when a prey sits down on the open trap. Mimosas, on being touched, fold up their paired leaflets, and this movement progressively affects neighboring parts of the plant. It has been found that stimulus conduction in plants, apart from being due to hormones, can also be accompanied by electric processes such as we know in the nerves of animals.

The principle is very likely the same in both cases, except that conduction is, as a rule, much slower in plants. Plants have no nervous pathways, nor do they have nerve centers for the correlation of sensory perceptions. In comparison with Man and animals plants are soulless creatures for all their graceful appearance.

III

ORGANISMS
AND ENVIRONMENT

---

1. *Adaptations*

Daedalus and Icarus

A Greek myth tells us about an inventive Athenian who went to the isle of Crete to evade punishment for murder. Minos, the king of Crete, kept him captive there. Finally he decided to escape by flying. He made two pairs of wings from feathers and wax, one for his son Icarus and one for himself, and fixed them to the arms. Exuberant Icarus went too near the sun, the wax melted, and the boy drowned. But Daedalus, so goes the tale, made a successful flight.

The age-old wish of Man to be able to fly has in reality never come true. Of course we do fly with the help of airplanes and spaceships, but the aviator himself cannot fly without mechanical means.

Birds, by contrast, are independent of technical equipment, of special fuel, of airports and landing strips. They lift themselves into the air with natural ease. Why then is it impossible for Man to fly with the help of wings? If it were only a question of suitable wings we could nowadays easily produce them. But this, of course, is not all. The bird has not only wings where we have arms, but there is hardly an organ in its body that is not modified for flight.

In flight the wings beat the air forcefully and often for hours on end. This is the task of the breast muscles that we all know and appreciate as "breast of the bird," and which are much more powerfully developed than the corresponding muscles in the non-flying mammals. These muscles are attached to the breastbone and, as they are so massively developed, the breastbone is enlarged and strengthened by a keel, a structure completely absent on the breastbone of Man, nor would we think of asking for "breast of rabbit."

We are out of breath when we run for a few minutes. Birds not only achieve a much higher velocity but they can keep up the race for hours, some migrating birds even for days, without resting and without getting out of breath. Such a performance has no equal among mammals. The lungs of birds are of a completely different design. They have large air sacs, enormous extensions of the lungs that, acting like bellows, help to ventilate them with an efficiency unequaled in mammals. Some of these air sacs are placed around the heart and embedded between the muscles used in flight. They prevent overheating in the same way as the radiator cools the engine in our car.

The frame of an airplane is made of light metal alloys. The frame of a bird, too, is of an astonishingly light build. The bones contain air instead of heavy marrow. Dogs do not care for bird bones. Extensions of the air sacs grow into the bones of the very young bird, and that is how the bones come to be filled with air. One can demonstrate that there really exists a connection between the lung space and the center of a bone: if we fit a glass tube into the severed windpipe of a dead bird and break off the upper wing bone, a candle placed in front of the broken bone can be extinguished by blowing into the windpipe.

Even organs like the digestive tract, which at first sight seem to have nothing to do with flight, are modified. For many years,

I kept a parakeet. We were intimate friends and observed each other with interest. It was astonishing to see how soon after a meal of cherries the red color appeared in the feces. It would take less than two hours. Later I found out that birds, with the help of an extremely short intestine and an extraordinarily quick and energetic digestion, rapidly get rid of all unnecessary ballast and keep their weight down in this way.

Thus the body of a bird is designed in every detail to suit its habits. We say it is adapted to a life in the air. Icarus, even if flying with the best of wings, would have crashed long before he could have reached the sunny heights, because of the weakness of his arm muscles, the weight of his bones, the load of digestive ballast in his intestine, because he would have run out of breath and for many other reasons. Man is simply not adapted to life in the air. Wherever we look in Nature we meet the same extraordinary adaptation of living creatures to their environment. In the interests of survival and perpetuation of the various species of animals and plants these adaptations appear purposeful in the highest degree. Whether or not they can be explained only by the assumption of a creative intelligence will be considered later. For the moment, let us study this phenomenon in some other manifestations. In birds we have seen the whole range of adaptations; next let us consider a single characteristic, animal coloration, and study it in a series of different animals.

Protective coloring in fish

We stand on the shore of a lake and watch a swarm of minnows, little fish the length of a finger, darting about in the shallows. Seen from a distance, they are hardly recognizable, so perfectly do they match the shade and color of the bottom of the lake. As we continue our walk we arrive at a stretch where, instead of being light sand and pebbles, the bottom is dark and muddy. There the swarms of minnows are dark like the ground beneath them. This seems to be a very useful protective device against predators in the water and above it. Or is it merely a coincidence?

We catch a few dark minnows and put them in a glass dish, with a light cloth underneath. They start turning visibly lighter and after a few minutes they have adapted themselves to the changed environment. If we set the glass dish on a dark cloth their bodies will turn dark again within a few minutes.

Occasionally a fish stands out spectacularly in a crowd of others, because it is dark against a light background. This makes us see more clearly how well the others are adapted. If we succeed in catching the odd fish we shall find that the lenses of its eyes are opaque or that it is blind for some other reason. But we do not have to wait for the chance of finding a blind fish to prove our point. If we temporarily cover the eyes of a minnow with an opaque mixture of soot and vaseline, the fish turns dark and is no longer able to adapt its color to the background. The change in color, an adaptation to the environment, is obviously stimulated by vision.

How can a fish change the color of its skin without changing the skin itself? Here the microscope enlightens us. Embedded in the skin of the fish lie innumerable little star-shaped cells, filled with fine-grained pigment. Nerve fibers lead from the pigment cells to the brain, and the arrangement of the pigment granules is controlled by nerve action. Under the influence of nervous excitation the black pigment collects into a spherical mass in the center of the cells. In this state the minute masses of pigment are invisible to the naked eye and the skin turns light. When the nervous stimulation stops, the pigment spreads out and fills all the branches of the cell (Fig. 55). The skin looks as if the pigment has been poured all over it and the fish turns black. The nervous pathways involved in the reaction are well known. If the responsible nerve at the tail end is severed by a pinprick, the tail turns black and remains incapable of adaptation. There are other colored pigments besides the black one. They can expand and contract and bring about satisfactory adaptation to the color pattern of the environment.

Fig. 55. Pigmented cell in a fish. Left, the pigmented cell expanded in a background of dark coloration; right, the cell retracted in a light background.

Flounders are masters in this sort of adaptation. In their early youth, while they still swim freely in the water, these extraordinary fish have a right and a left eye, being bilaterally symmetrical like most other animals. Later, however, when they come to lie on the bottom of the sea, the eye from one side moves across to the other side, and the flank of their body on which they lie becomes white like the belly of free-swimming fishes. The flank facing upward becomes not only dark or light but reddish, yellowish, greenish, or bluish, deceptively matching not just the color but the pattern of the background. Above a uniform sandy background the fish shows a uniform patterning; against pebbles, with their patchy distribution of light and shade, the flounder too shows a pattern, in which the size of the speckles corresponds most astonishingly to the size of the pebbles.

The example of color changes in fish was chosen with deliberate intent: not only is the adaptation perfect but it is based on the existence of an elaborate mechanism of color cells and nervous fibers, obviously designed for adaptation to the environment; it is impossible to regard them as accidental and without biological significance. The value of protective coloring has been experimentally proved; we have statistical evidence that fish badly adapted to their environment become easy victims to birds of prey.

The colored kaleidoscope of anger and love

Amphibians, reptiles, and even invertebrates are able to respond to environmental changes of color and light. It is well known that the tree frog, whose green coat makes it practically invisible among foliage, turns a grayish brown when sitting on bark. The color change of the chameleon is proverbial. The fame of this animal derives from the striking display of colors with which it responds to irritation. Similarly the octopus dramatically mirrors its emotions in its skin by making it flash in a variety of colors. In the beautiful marine-biological aquarium at Naples, I once observed a gurnard, one of those peculiar fish that can taste with the tips of their fins, and found that its belly and sides turned suddenly pale if I shook a finger at it. After half a minute the normal blood-red skin color returned, but the slightest movement in front of its basin made it turn pale again, while the animal itself did not move from the spot.

If in these cases color change is merely an expression of emotion, comparable to the blushing or blanching of a human face, it may also acquire biological importance. Thus, during the mating season, the male of some fish display a spectacular show before the female. A mechanism originally acquired for protective purposes is here put to a totally different use.

How to be invisible

Animals use colors chiefly for protection. To the thoughtful observer, Nature appears here as a creative artist with unflagging inspiration, imagining and realizing all sorts of variations. Insects, for instance, cannot change their color as quickly as some lower vertebrates, so they have to blend permanently with their natural surroundings. Green caterpillars, with their remarkable colored patterns, astonishingly resemble their environment. Grasshoppers are grass green. Their relatives, the locusts, who are desert dwellers, are yellowish gray; they share this coloring with the mice, birds, and other animals of the same habitat. One spider is perfectly disguised when lying in wait on certain flowers for

Fig. 56. Caterpillar in an attitude of defense. Note how the animal becomes part of the environment, a seemingly protective device.

its insect victims. It can change the color and pattern of its body when lurking on different flowers. Another spider resembles the lichen-covered tree trunk so closely that even the sharp eye of a bird cannot detect it. It sits still during the day but goes hunting at dusk.

Some crabs have no protective coloring of their own: they collect bits of plant material and fix them to hooks on their chitinous armor, so that their body disappears under this cloak and cannot be distinguished from the surrounding vegetation. If they are taken to some other setting where their disguise is no longer effective, they quickly pull it off and change it for a more suitable one.

Cuttlefish protect themselves with a cloak of a different kind. They have a gland that produces the well-known sepia ink. When they are chased they hide in a cloud of black ink. The animal effectively combines this black magic with a vivid color change. When pursued it first turns dark. Then it expels its black cloud of ink, which floats through the water, a duplicate in size and color of the darkened animal. But meanwhile it has turned snow-white. The pursuer, not expecting such a change, goes after the cloud of ink, while the anticipated morsel escapes in its different make-up. Numerous free-floating aquatic animals use the simplest method of being invisible: they are entirely transparent.

Birds and mammals, whose feathers and hair coat are made of horny tissue, can change color only by producing a new vestment. Weasels, snow hares, and snow hens seasonally change color. In autumn their brown summer dress is exchanged for a white winter coat. Polar bears, polar foxes, and other inhabitants of the subpolar regions have a permanently white coat; thus they fit in with the eternal snow.

Protective colors often become fully effective only when combined with appropriate instinctive behavior. Certain looping caterpillars look exactly like a dry twig (Fig. 56) provided they do not move. They go foraging only in the dark. Other animals with protective coloring feign death, when in danger. Complete immobility, however, is not always the best way to adapt to environment.

If a bittern is suspicious of an approaching animal it stands rigidly still among the reeds, with its neck and beak stretched upward. Its yellow underside, barred with vertical dark stripes, is invariably turned toward the potential foe. On a calm day, the rigid bird resembles the still reeds around it. On a windy day it sways with the rhythm of the reeds.

Warning coloring

Not every coat is a protective one. The black raven in a snow field is as conspicuous as a red deer in a green meadow. They do not need to be camouflaged because they are protected by their size or speed and themselves are not predators attempting to stalk their prey unnoticed.

Small animals sometimes show very conspicuous colors, although one would think they would need the protection of camouflage. The clumsily moving salamander attracts attention by its yellow and black pattern. It does not seem to be in a hurry and looks quite confident. Indeed, it is protected against most aggressors because its skin produces a nauseating slime, which is poisonous to small animals and can induce vomiting in dogs. One unpleasant encounter with this animal, and its garish colors will be remembered and shunned forever after. The combination of striking coloring and a poisonous secretion is frequently found in insects also. Thus color is equally effective as a warning in some cases and as a camouflage in other cases.

The secrets of the deep sea

Camouflage makes sense only in bright daylight. Animals that live constantly in the darkness of caves, such as the amphibian Proteus, which lives in subterranean rivers under limestone mountains, have no adaptive color. Neither does it occur in the unfathomed regions of the deep sea, into which the sunlight never penetrates. But there we find adaptations of a different kind and more extraordinary than any we are familiar with on land.

The summit of the highest mountain on earth, Mount Everest, is 29,141 feet above sea level. The bottom of the sea is about 36,198 feet below the surface of the sea at its deepest. The average depth of the ocean is 13,130 feet. The rays of sunlight striking the sea are quickly absorbed, even where the water is quite clear. At a depth of 1,500 to 1,800 feet the last remnants of light have gone. As the photosynthesizing plants that are the source of all food are dependent on light, the lightless depths of the sea are barren. Organisms that live in such depths depend on organic substances produced by the plants floating in the illuminated surface waters. There is a constant rain of the dead bodies of such plants and of small animals sinking down into the depths of the ocean. For a long time the greatest depths of the oceans were considered uninhabitable. Who would care to people these regions of total barrenness and darkness? But one had underestimated the abundance of this organic rain and forgotten to take into account the fierce struggle for existence, which forces animals to exploit all possibilities – not to mention the creative power of Nature, which equipped them for the conquest of seemingly uninhabitable territories with the aid of special adaptations in form and function.

When, about a hundred years ago, a research vessel lowered its nets for the first time to deep-sea levels, the explorers were astonished to find these populated with a most varied assortment of fish, cuttlefish, crustaceans, and all kinds of other organisms.

The most astonishing fact was that most of them had well-developed eyes, quite in contrast to the blind inhabitants of caves. The explanation was soon found: their bodies are often completely covered with luminous organs similar to those of glow-worms. They light their own little lamps to pierce the eternal darkness of the abyss. When the deep-sea fish were studied more closely, it became clear that these luminous organs have mainly three kinds of functions.

Often large luminous organs are found near the eyes. Like searchlights they send out a beam of light in the direction in which the eyes are looking. Floodlights are equipped with reflecting mirrors, and so are these organs. Behind the luminescent tissue a reflector of microscopically small crystals enhances their efficiency. These living lamps can be blacked out, either because the lighting process can be turned on or off at will, or by turning them inward with the help of special muscles.

Besides, there are usually great numbers of smaller luminous organs distributed over the body. Their arrangement differs in different species. They are the distinguishing features of a given species and correspond to the characteristics such as shape, coloring, and pattern by which land animals recognize their kin.

The most extraordinary luminous organs found in deep-sea fish hang like light bulbs from the tip of the elongated first ray of the dorsal fin. Like a fishing rod, this organ is trailed along right in front of the voracious mouth of the fish, and the fate of those that let themselves be attracted by this luminous bait can easily be imagined.

What a different world! Water right and left, water above and below, water fore and aft, water in limitless quantity surrounds the floating population. Darkest night reigns, even when for us the sun stands at its highest. And yet there is light, light of a special kind, breaking the monotony of these fastnesses, producing a veritable riot of illumination. Our nets bring up only a small number of the denizens of the deep, because only a few let themselves be caught. So we know only a meager sample of what really lives there.

The American naturalist William Beebe was one of the first to become aware of this fabulous life. Descending in his bathysphere, he saw through its quartz windows what no human eye had ever seen before. Big or small, all creatures of the deep had their own little lamps arranged in various ways. Small fish and large ones, sometimes several yards long, flitted past the windows. Prawns swam past and when frightened jettisoned clouds of a bright luminous substance, behind which they retreated into darkness, the reverse of the sepia clouds squirted out by the cuttlefish in the light upper regions. Beebe reached a depth of 1,870 feet. This was on August 15, 1934. Since then Auguste Piccard has constructed a little submarine, with delicately adjusted buoyancy, independent of cables. On January 23, 1960, this bathyscaphe, manned by two persons, descended freely to the bottom of the Pacific Ocean, reached a depth of 35,958 feet, and returned to the surface unscathed. All the way down the observers saw living animals.

It takes courage to undertake such investigations. Man is as little adapted to the terrific pressure at such depths as the animals living there are to the conditions in the surface waters, where they quickly perish.

Physically, Man shows very few striking special adaptations. But his intellect has broken through all barriers set him by Nature. He is not fast enough to catch the fleet-footed deer, yet he reaches it with his bullets. He is not adapted to life in higher latitudes, and yet he spreads out into them by making himself clothes and by heating his living quarters. His body is not made for flying, but he builds himself airplanes and spaceships. He cannot dive down into the depths of the ocean, but he descends in a bathysphere. His intelligence reigns supreme and replaces the adaptations his body lacks.

2. Animal Migrations

When the warming sun of spring awakens new insect life in woods and meadows innumerable birds find a full table, and one day the swallows, our well-loved summer guests, are here again. They build their nests, bring up their young, and behave as if they feel very much at home and want to stay for good. But in fall, when the days get shorter and the nights chillier and food runs low, the swallows collect in great flocks and finally fly away to the south to find food and warmth.

There are parts of this earth where conditions never change all the year round. The deep sea is one of them. In most other places there is seasonal change; the tropics have their rainy and dry seasons, in our latitudes we have summer and winter. There are, of course, animals adapted to changing conditions, for instance nonmigrating birds. We do enjoy their presence during winter. But not all have been equipped by Nature to endure hungry days. This is the main reason for animal migration, of which bird migration is the most impressive. It is a most wonderful sight to encounter hundreds and thousands of birds along their favorite routes, as they steer a set course toward their winter quarters. What is their destination and what is their guiding star?

The mystery of bird migration

Bird watchers have thought of means to discover the destination and routes of the migratory birds. They catch them during their flight, or take their young out of the nest, and fasten a light aluminum ring to their leg. The ring bears a number and the name of the experimental station. This light burden is no hindrance to the bird. The bird is then released or put back into its nest. If later on the ringed bird is caught or shot en route or at its winter quarters, the ring may be noticed and sent back to the station, with a note to say when and where the bird was caught or shot. In this way it was found out that a stork ringed in July in the Baltic spent December of the same year in South Africa. Birds caught during migration yield information about routes of flight.

With the help of ringed storks it was established, after a number of years of observation, that storks breeding in Eastern Germany fly to Africa via Asia Minor and Israel, while storks from the west of Germany fly southwest and cross the Mediterranean at Gibraltar (Fig. 57).

Other migratory birds cross wide stretches of open sea. The American golden plover crosses the Atlantic between Nova Scotia and South America in an uninterrupted flight of about forty-eight hours. Most birds take more time over it: they dawdle here and there, conditions permitting, and may take weeks or even months for their migrations.

The longest route is flown by arctic terns, which during our summer breed in the northernmost parts of Europe and America, and migrate at the beginning of winter to the shores of the south-polar sea and to the antarctic continent. It has been calculated that they manage to fly a stretch of at least ten thousand miles twice a year.

Even more astonishing than the power to fly such great distances is the incredible sense of orientation shown by some species of petrel, which live during summer around the coasts of Britain and fly to a small island of the Tristan da Cunha group in the southern Atlantic Ocean as soon as summer begins in the southern hemisphere. What wonderful sense of orientation guides the bird to a rock in the middle of the ocean? Quite a few migratory birds do lose their way in gales and drown. Some get exhausted and rest on ships or as lost pilgrims reach islands they normally never visit. Yet the majority reach their destination. How do they find it?

Birds have very good eyes—better than ours. In flight they scan larger areas than earth-bound creatures. This excellent sense of vision was thought to explain the riddle of bird migration. We do know that birds steer by landmarks. As in most species the young and old migrate together, knowledge of the migratory route would

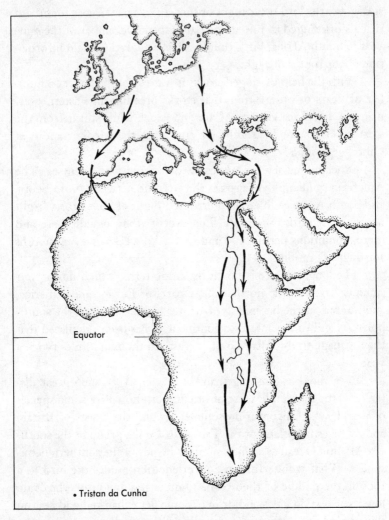

Fig. 57. The migration route of the stork. Note the different routes of storks breeding in East and West Germany.

be handed on from generation to generation. But this is not the whole explanation.

In birds such as starlings or crows, the young depart before the old, and in the case of certain kinds of gulls they leave after the old.

They are therefore without experienced leaders and the open ocean offers no landmarks.

Some new evidence was obtained by an experiment conducted by ornithologists. In East Prussia young storks were taken out of their nests and locked up until all the old storks had left Germany for Africa. The young ringed storks were then moved to Western Germany and set free. Would they, without guidance, set out on a flight to Africa? Would they fly via Gibraltar, like all storks from Western Germany, or would they fly via Asia Minor, like their parents and the other storks from the same eastern birthplace? Neither happened. They left in a southeasterly direction, the very direction they would have had to take to find their traditional route via Asia Minor. Thus they left their unfamiliar starting place, traversed the Alps, and were last seen in Italy. The experiment shows that the knowledge of the direction of their migratory route is innate in storks. But this does not explain how the right direction is recognized. More recent experiments on starlings and other birds have shown that they use the sun as compass, taking account of its course to calculate their whereabouts. On the other hand there are birds, like the warblers, that migrate at nighttime only. Like the mariners of old, they steer by the constellations. In an experiment conducted in a planetarium, the birds set off for the south even under the artificial sky. Gradually this sky was changed in the way the birds would see it during their journey south, and this sustained their southward urge. When at last the stars of the African winter sky were shown, the birds came to rest. As far as they were concerned they had reached their destination.

But all this does not as yet fully explain the riddles of bird migration. A bluethroat caught and ringed in Potsdam in Prussia during the breeding season and released in Lower Bavaria to everyone's astonishment returned to Potsdam within a few weeks. Swallows and starlings, removed many hundreds of miles from their nesting places, found their way home again within a few days, although they could not know where they had been taken.

Extensive migrations are known not only among birds. Periodic lack of food can bring on seasonal migrations in other animals, for instance in reindeer. Sometimes the migrations are irregular, sometimes they happen just once as the consequence of overbreeding or lack of food. The dreaded swarms of locusts and the fearful progress of wandering rats are such cases. Occasionally animals make long journeys to their traditional breeding places even though they have settled down to live far from them. The eels do this. Their life history was an unsolved puzzle until the Danish biologist J. Schmidt was able to suggest a possible explanation.

The migration of eels

The young eels (elvers), which look like tiny earthworms, ascend in uncountable quantities from the sea into the rivers on the coasts of the Atlantic, the Mediterranean, the North Sea, and the Baltic. They swim up the rivers, overcoming difficult obstacles like slippery rocks with astonishing agility, and finally reach the headwaters of rivers and brooks. There they stay for six to ten years, feeding and growing to a length of five feet and reaching a weight of ten pounds.

But never yet has an eel laid its eggs in fresh water. When the eels are fully grown an urge to migrate makes them swim in the opposite direction, back to the sea. There they disappear and are never seen again. It is only a few years since we have known that eels from the Mediterranean migrate through the Straits of Gibraltar, and Baltic eels via the North Sea, and that they all cross almost the whole of the Atlantic Ocean in order to mate and to deposit their eggs in the Sargasso Sea. This happens at a depth of over a thousand fathoms. The young, the larvae hatching from the eggs, are flattened and do not look like eels at all. Their development is a slow process, during which they drift gradually toward the European coasts to start their upriver migration in their fourth year after having changed into young eels.

How was this found out? What was needed was perseverance, patience, and money. One knew for a long time that the transparent leaf-like fish are the larvae of the eel, because they had been observed to change into eels in captivity. And then the Danish biologist Schmidt sailed the ocean for years fishing for eel larvae. His plan was to follow the direction in which eel larvae get smaller and smaller. He thus eventually reached their breeding place and even managed to fish their eggs from the depth of the sea. This breeding place is almost identical to that used by the eels coming from the American seaboard. The offspring of the two eel populations then migrate away in opposite directions. Whether their parents die or whether they live on in the depth of the sea we do not know. It must be said that this story has recently been challenged and that the last word on eels has not yet been spoken.

The faithful salmon

The eels spends its youth, the time of its most vigorous growth, in rivers, and then goes on a trip to the sea. The salmon does it the other way round. It leaves its cradle, the river, as a small, inconspicuous youngster and makes for the ocean, where it develops and keeps an excellent appetite. When it returns into the river several years later to deposit its eggs there, it has grown into a coveted haul for fishermen.

During its long migrations the salmon shows remarkable strength and perseverence. Swirling whirlpools and thundering waterfalls cannot impede its progress. Propelled by the powerful beat of its tail, it leaps where it cannot swim, until it eventually reaches the upper river and its tributaries, often high up in the mountains. There, in a pebbly shallow place, the female scoops out a hollow. After spawning the trip downriver is rather easier, but quite a few of the spent fish perish. Those that reach the sea start on another bout of feeding, and such survivors as there are migrate, year after year, back to the rivers to spawn.

In many once famous salmon rivers the number of fish has declined. This is due to the pollution of rivers by factory wastes and to the construction of weirs and artificial dams over which salmon cannot leap.

For a long time it was suspected that salmon ascend from the sea to the river of their origin and the expert distinguishes salmon of different rivers as belonging to different breeds. This would hardly be possible if salmon bred indiscriminately.

It was thought they might stay near the mouth of the rivers of their origins. But this is not so. Just as one can band migrating birds, so one can attach small numbered plastic or metal disks to the fins of fish. In this way, biologists have marked salmon, particularly the Chinook, and have authenticated its remarkable migration.

The Chinook, or the King salmon, ranges from California to the Bering Straits. The young fry are hatched in the fresh-water streams of the Pacific Northwest. Within a year after hatching, the salmon instinctively heads out to sea. It runs the gantlet of its enemies—mink, raccoon, bear, duck, heron take their toll of the young salmon and so, alas, do the polluted waters where factories and cities send their waste into the streams. Only a very small proportion of the salmon actually reach the sea (of 5,000 eggs laid only about 50 hatch), where food is plentiful. Some of the young salmon swim out only a few miles into the sea; others travel hundreds of miles.

After four to six years the salmon begins to migrate to its spawning grounds; the spawning ground is *not* the sea but the kind of stream in which the salmon was hatched. It is known that some salmon return to the very stream in which they were hatched; others find their way into similar waters.

Nevertheless, the adult salmon take several months to migrate from the sea to lay their eggs in the kind of water in which they were spawned. They perform amazing feats to do this, sometimes leaping rapids nine or ten feet high. Man has helped in this; he has constructed "fish ladders"—artificial sloping waterfalls—to make it easier for the salmon to "run" upstream.

During these runs vast numbers of the salmon are caught (in nets and traps) and canned or smoked. But enough reach the spawning ground to lay their 10,000 or so eggs; enough males are left to spread their milt (sperm) over the eggs. Then the adults, having carried out their act of reproduction, seem to be aware that life is over for them. They change color, the skin becomes very slimy, the flesh pale. Slowly they begin to float downstream—tail forward. In a few days they are dead and are washed ashore to furnish food for the ever-hungry bear or raccoon.

---

3. Symbiosis and Parasitism

The hermit crab and the sea anemone

The bather on a rocky shore who pays attention not only to the beauty of sea and coastline, but also to plants and animals between tides, will come across some sea shells that move about with astonishing speed. On closer inspection he will notice that, instead of the head and foot of a sluggish whelk, a pair of curious eyes are looking out of the shell. They belong to a crab, as do a number of agile legs moving in and out of the entrance. He has found a so-called hermit crab. In contrast to its many well-armored relatives, this crab has a soft abdomen that would leave it unprotected against enemy assault. Therefore, it protects its abdomen in an empty shell, which it carries about all the time. If it cannot find an empty shell it "requisitions" an occupied one. Short shrift is made of the lawful occupant, and with one stroke there are gained both a desirable property and a good meal.

This is plain assault and murder, which is a crime among mankind, but the usual routine in the animal world, with its law of the jungle. It seems all the more pleasant to find that some of these animal aggressors can also enter into a lifelong peaceful alliance with another creature. Although the basis for such an association is not ethical principles, the fact is none the less remarkable. Certain species of hermit crabs, for example, regularly carry on top of their house one or several sea anemones. These are primitive animals, colorful as flowers. Usually they attach themselves firmly onto stones in the tidal zone. Under their deceptively harmless disguise they act like ruthless cutthroats. Their tender tentacles carry thousands of microscopically small poison capsules that explode at the slightest touch, like those of their relatives, the free-swimming jellyfish, the sting of which is known to many a bather. The poison capsules are used against fishes and other animals, which die when they come into contact with the tentacles and can then be swallowed at leisure by the anemone. One might wonder whether the sea anemones on the hermit crab had got there by mistake. Patient observation teaches us differently. When the crab grows, its abode becomes too small for it. It finds a bigger shell and moves into it. When it does this it takes the sea anemones with it too. With its pincers, otherwise weapons formidable to any prey, it gently loosens the foot of its anemone from the old house and presses its new one against it, until the anemone attaches itself. Thus ends a successful move. Sea anemones, usually so touchy, do not defend themselves in this case.

We are dealing here not with an accidental encounter but with a deliberate association, of advantage to both partners. The hermit crab is protected by the poisonous capsules of the sea anemone, especially against attacks from its most deadly enemy, the octopus. The sea anemone in turn is carried about by the hunting crab, and when the crab feeds, cutting up its prey with great speed, many a dinner falls on the anemone's tentacles.

Another species of hermit crab shows a still more intimate relationship with its anemone. It is never without one, and it is quite a special species of anemone, which has never been found anywhere except on the house of that particular species of crab. There it sits protecting the entrance with a crown of tentacles covered with stinging capsules. The tentacles are held in a plate-like fashion under the mouth of the crab, ever ready to catch the leftovers from the crab's meal.

Ants and plant lice

When two live together for each other's benefit one speaks of them as living in symbiosis. This does not mean that the partners have always to live as close together as the crab and the sea anemone; the connection can be a loose one. We find an example of this in our garden.

On roses, elder bushes, and many other plants there can often be found dense clusters of plant lice. The gardener hates to see them, because they live on the sap of plants, sucking it through a tubular pointed part of their mouth. Their excretion is rich in sugar. This waste of valuable food is explained as follows: aphids do not move about a lot and their need for fuel is therefore relatively small. But they need a lot of protein for building up protoplasm, because they grow quickly and reproduce rapidly. Plant sap is as poor in proteins as it is rich in carbohydrates. The aphids therefore have to take up a lot of sap in order to cover their protein supply, at the same time taking in more carbohydrates than they need. This surplus is excreted as sugar.

Sugar juice is a sticky fluid. So as not to soil themselves with it, the plant lice have the habit of squirting it out in droplets, which cover the foliage as so-called honeydew. Bees collect this without taking any notice of the suppliers. Ants, on the other hand, which also appreciate the sugar, have attached themselves to certain species of plant lice. They fetch the sugar straight from the source, namely from the anus of the aphids. In these plant lice the excretion of the sugar is delayed and held back until an ant drums at the aphid's abdomen with its antennae. This leads to the gentle release of a droplet into a container made by a special ring of hairs around the aphid's anus. From this the ant takes the sap as from a cup. Ants, which are the fiercest enemies of other insects, which they hunt and drag into their nests, tend and protect the aphids because of that sugary honeydew. In autumn some species of ants carry the eggs of plant lice into their nests, in order to take the hatching young back to their food plants in spring, after safe hibernation.

Symbiosis in plants

So far we have seen examples in which two different animals

live together to their mutual benefit. This phenomenon is, however, not restricted to the animal kingdom. The lichens, which
we can find everywhere on trees or rocks, are a community of
two plant organisms: the hyphae of fungi and the cells of algae,
supplementing each other's needs. Similarly, the strange knots or
nodules on the roots of peas and other leguminous plants are
nothing but the seat of certain bacteria, which always live there
and can make direct use of the nitrogen in the air, contained in
the surface layer of the soil. The root of the pea plant, harboring
the bacteria, in return gets the surplus of nitrogen acquired by
them. Thus peas can thrive in soil that is lacking in fertilizer and
its nitrogen compounds. Farmers make use of this by rotating
their crops, growing alternately grains which are nitrogen con-
sumers and legumes which are nitrogen producers. An animal
and a plant can also live in symbiosis. The relationship between
flowers and insects is an example of such an association on a
grand scale.

Flowers secrete a sweet juice, the nectar; this they offer as
food to insects. Colored petals are striking advertisements and a
cloud of scent helps the guest to find the restaurant. All these
arrangements are not only useful to the guests but to the plant
also, because during their foraging flights the insects carry pollen
from one flower to the next, bringing about the setting of seeds.

This example can hardly be called a case of symbiosis, because
insects and flowering plants, despite the far-reaching consequences of their encounters, live their own very independent
existences. Relationships of course can be of all degrees of close-
ness, sometimes extremely close indeed.

Intimate associations

Southern European countries and especially the tropics are the
homes of termites, or white ants. They are actually no near relatives of the ants but they lead a similar social way of life. Ants
can become very disagreeable when they invade our houses, but
they are harmless compared with termites. Unless controlled, ter-

mites can cause great destruction by chewing up the beams of houses from within.

It is extraordinary that animals should like the taste of dry wood and be able to live on it. As a rule wood and cellulose are resistant to the digestive juices of most animals, and in this respect the termites are no exception. Their intestine, however, contains single-celled animals, flagellates, which, with the collaboration of bacteria, can split cellulose into sugar. The flagellates find in the intestine of the termite shelter and food in plenty and they produce on their part a surplus of sugar that guarantees a sufficient supply for their host.

An American biologist, Dr. Cleveland, has been able to explain and demonstrate this interdependence. The flagellates are more susceptible to high temperatures than the termites. If one therefore puts the termites into a heated chamber, the intestinal guests are killed before the termites are affected by the heat. But death by starvation is their fate, even if they are offered plenty of wood. The reason is that they cannot digest it. If one gives them a chance to swallow a new supply of flagellates in time their digestion begins to work again and they stay alive.

A similar relationship exists between ruminants and cellulose-digesting bacteria living in their stomachs.

There exist still closer relationships in which the symbiont does not just live in a cavity of an organ but within the very tissue cells themselves. The fresh-water polyp, a distant relative of the sea anemone which we discussed at the beginning of this chapter, is only about one eighth of an inch long and catches its food with tentacles covered with poison capsules. The food consists of water fleas and similar prey larger than the polyp itself. Most fresh-water polyps are brown, but there exists a vividly green species, the green showing all the properties of chlorophyll, the pigment in the foliage of plants. It was discovered that the polyp owes its color to the presence of thousands of single-celled algae, plants that thrive and reproduce within the cells of the polyp's intestine. This too is a case of symbiosis to the mutual advantage of the partners. The algae live safely within the armed polyp and receive at first hand the end products of the animal's metabolism, namely simple nitrogenous compounds and carbon dioxide, from which, in the manner of plants, they build up organic compounds of their own. The polyp, on the other hand, is said to feed on the surplus of such compounds, and it can definitely make use of the oxygen liberated by the plant's metabolism. Here, within the body cells of an animal we have a small-scale model of the great cycle of interdependence that exists between animals and plants as a whole. In fact the green Hydra can withstand the lack of oxygen in the water and lack of food much more easily than its brown relatives. When the green polyp produces eggs, algae wander from its body into the egg cells so that the new animals are born with their symbionts to sustain them during their lifetime as a kind of inheritance.

Use and abuse

The most pleasant relations among animals or human beings are those in which two work together in such a way that what is useful to one is of benefit to the other. But in the best of cases this can turn out to be a precarious relationship, which might easily change from mutual help to one-sided exploitation. This has probably happened quite frequently, and here and there we seem to be able to trace the steps by which events took such a turn.

In the sea we find crabs that carry anemones attached to their broad backs. On the other hand, there are crabs that hold their sea anemones in their strong claws and push them into the face of an approaching enemy. If the anemones do not stand this treatment very well, the crab plucks some fresh ones from the colored multitude that live at the bottom of the sea. This relationship can no longer be called symbiosis; it is brutal force used by the stronger in the exploitation of the weaker animal.

Even the delightful and peaceful relationship between flowers and their insect visitors is occasionally marred by imperfection. Some flowers, such as the wild sage, have their nectaries at the base of a long corollaceous tube and are therefore suited to be most effectively visited by bumblebees with long sucking tubes. When the bumblebees press their long proboscis into the flower, they brush past the stamens and the stigma and bring about fertilization. But there are other bumblebees with short proboscises, which bite their way through the side of the flower to help themselves to the nectar, without giving any return service.

When one symbiont lives within the body of the other there is all the more need for the maintenance of a balanced relationship. The host might begin to suppress the lodger, or the lodger might misbehave, spreading unchecked and turning from a guest into a germ of disease, from a symbiont into a parasite. It is not always easy to judge such mutual relationships correctly.

Parasites

A true parasite cannot hide its crime and has telltale features that give it away. It is amazing how a way of life can change the whole make-up of an organism. This is most convincingly demonstrated by the comparison of two related species, one free-living and self-supporting, the other a parasite.

In our ponds and lakes as well as in the sea live enormous numbers of small crustaceans, though very few people know about their presence. The clear water of a mountain lake looks quite unpopulated, even to the careful observer. If, however, one draws a very fine net through the water and rinses its contents out into a jar, one realizes that all over the lake lives a most varied community of creatures. A hopping and jumping is going on in our jar, as if a sac of fleas had been emptied into it. A large proportion of these animals, which measure hardly one twenty-fifth of an inch, belong to a group of crustaceans known as copepods. They swim jerkily about with lively little legs. Their food consists of microscopically small particles, which they filter from the water with the help of tiny bristles. Their long feelers are covered with organs for tasting and smelling, and they find their way from darkness to light with a simple tiny eye. The females carry

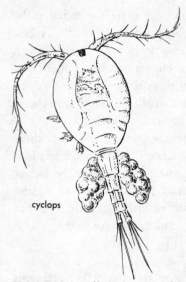

cyclops

Fig. 58. A small crustacean (a copepod) bearing two egg sacs. (Enlarged about 50 times)

packages of eggs about with them until the young hatch — if it ever comes to that (Fig. 58). Millions of these small crustaceans finish their life in the stomachs of fishes, many of which fancy them as food at some time or other.

Some species of these crustaceans have turned the tables and live on fishes, settling down on skin or gills as parasites. Only the specialist, after a detailed study of their shape and life history, recognizes them as relatives of the free-living copepods (Fig. 59). The layman would probably call them worms. The segmented body and lively legs have disappeared. They have given up swimming and let themselves be carried about by their host. Their jaws have changed into suckers with which they feed on the body fluids of the fish. There are no more sense organs, not even an eye, because having settled down for life near a sustaining medium they have no further need to search for food. Living in abundance, they grow larger than their free-living relatives. There is only one drawback in this ideal situation: if they did reproduce on their host, if the young settled down beside the old, the fish would soon be exhausted and with the death of their victim the parasites would have deprived themselves of their food supply. Therefore nature has decreed that the young

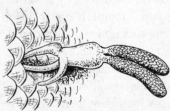

Fig. 59. A small parasitic crustacean (on a fish) — a relative of the crustacean in Fig. 58 — bearing two rear egg sacs. (Enlarged about 5 times)

should move away and seek to find a fish of their own to settle down on. This is not at all easy for such small animals, although at this stage they still have sense organs and legs. Uncountable numbers are unsuccessful and perish. Yet the species survives. The parasites feed so well that they can produce lots of eggs, many more that their free-living relatives, and thus one or the other of the thousands of young manages to establish itself and to live on yet another fish.

Parasites that live inside their hosts show still greater changes in form.

There exists the fraternity of intestinal worms that plague birds and mammals and that occur even in Man, despite all precautions of hygiene. Among them, the tapeworms have the most striking parasitic features. As long as there are not too many of them they do not endanger the life of their host.

One species of tapeworm in Man can reach a length of several yards. Its head is diminutive. Sense organs and nervous system are primitive, with hardly so much as an indication of a brain. The head carries suckers to attach itself to the intestinal wall and is therefore not easily voided with the feces. It does not suck food from the wall of the intestine but lives on the intestinal contents. As the host digests the food the tapeworm needs neither mouth nor gut. It absorbs through the surface of its body the nourishing fluids that ought to reach the host.

It produces millions of eggs during its lifetime and this is essential if the species is to survive. The life history of an internal parasite is very involved. In the tapeworm under discussion the eggs within the intestine of the host finally reach a body of water. From them emerge ciliated larvae (a in Fig. 60), which must find their way into little crustaceans, using them as intermediate hosts (b); for the tapeworm's further development it is necessary that a fish swallow a crustacean with such a tapeworm larva in it. The larva has to find its way into the muscles of its second host to encyst itself in them. Only if Man eats such a fish raw or insufficiently cooked does the tapeworm reach its ultimate goal and come to maturity in the human intestine (Fig. 60c). One may

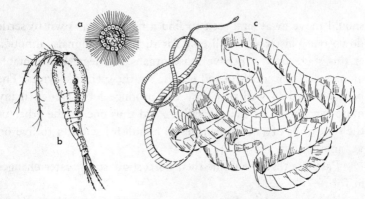

Fig. 60. Life history of a tapeworm: (a) the ciliated larva; (b) the larvae penetrate and live inside a small crustacean; (c) an adult tapeworm, Bothriocephalus, as found in the intestine of Man.

well ask why the digestive juices of the stomach and intestine work on the meat of the fish but not on the tapeworm too. It is just one of the many adaptations to its mode of life inside an intestine that the tapeworm's tissues are chemically equipped to withstand digestion.

These pale creatures are not very pleasant company. They can become troublesome not only because of the quantity of food they consume at the expense of their host, but more so because they produce poisonous substances. However, if we compare them with other parasites they are relatively harmless despite their size. We know microscopically small bacteria and others at the borderline of visibility that are the cause of much more dangerous diseases. The degree of damage is not proportional to the size of the parasite but depends on the nature of the poisonous metabolites that it produces.

If, however, a parasite is so big as to eat the host then it turns into a beast of prey. A leech that sucks the blood of a horse is a parasite; when it sucks empty a snail and kills it, it is a beast of prey. A sharp line of demarcation exists here as little as at the other end of the scale, when we deal with the question of good-neighborliness in symbiosis.

4. *Herds and Animal Societies*

Animal aggregations that are not true societies

A flock of vultures collecting around a corpse is an assembly brought together by the fleeting chance of a good feed. Similarly, communal sleeping or nesting places can bring together temporarily many members of a species.

A proverb says: Opportunity maketh the thief. On the other hand opportunity can make for mutual help and friendly cooperation. Where many sea birds nest together on an island the warning cry of one animal rouses the whole community, and if an enemy approaches they take determined and concerted action, confident in their sheer force of numbers. Chamois, which live in herds, or marmots, which build their warrens close together, benefit similarly. Many pairs of eyes see more than one pair, and where many noses are lifted to sniff the air the animal with the most acute sense of smell will raise the alarm and send them all fleeing. In winter wolves hunt in packs, some chasing the prey and others trying to cut off its retreat by surrounding it and closing in on it. As the seasons get more hospitable and dire hunger does not hold them together, each wolf goes its separate way.

The ties that hold crowds of monkeys together are closer, because they live in families. Frequently several monkey families live together, the strongest and most experienced male being the guard of the group and the leader on their foraging expeditions. But with the more closely knit social unit we find also its shortcomings: constant quarreling and fighting among its members.

Family life and the association of several families for communal enterprise and defense are no doubt the origin of human

society. But much older than any human society are insect socie-
ties of bees, wasps, ants, and termites. The essential feature of a
society is not the number of its members but its inner organization, the division of labor and the subordination of the individual
for the benefit of the community.

In this respect the beehive is a most interesting example of a
society in which everything that happens serves the community
as a whole and never the selfish ends of the individual bee. The
only limitation to our admiration is the fact that this social unit
is governed not by an idea consciously pursued but by inborn
urges manifesting themselves in rigidly fixed behavior patterns.

The beehive

A community of honeybees is not a group of families but one
single large family, counting about 80,000 heads. Bees used to
build their hive into hollow tree trunks, which were not as rare in
the past as they are in our days of the cultivated forest. This
would have meant an acute housing shortage for bees if we had
not turned them into domestic animals for the sake of their honey.
Now we give them wooden boxes or straw baskets to live in.

In every hive lives the mother of the family, the queen bee,
with her 40,000 to 80,000 children. The total number of her progeny is several times larger. But most of them die after a few
weeks or months, while the queen lives on and is fertile for four
to five years. She is not weighed down by affairs of state. Her
most important duty is to lay eggs. This is a full-time job, because an active queen daily produces 1,500 eggs; this means that
day and night she lays an average of one egg per minute. The
other bees are not involved in this. The so-called worker is a sterile female with nonfunctioning ovaries and never lays eggs under
normal conditions. But the workers excel in all other female
virtues — domestic tidiness, care for the larder, and the tending of
the brood. In spring big-headed drones turn up in the colonies of
bees. These are male bees and their virility is their only function. In summer when they have become useless, the workers

begin to resent the presence of the drones as if they were parasites in the hive. This leads to the dramatic battle of the drones, when the males are mauled and stung and chased out of the hive to die an inglorious death. They are defenseless against assaults of the workers, since they have no sting and do not show much fighting spirit. They do not know how to find food and are doomed to starve at the very entrance of the hive.

To the queen and to the workers (both females) are allotted different tasks because their bodies are developed differently. Moreover, there is a well-organized division of labor among workers. In contrast to Man, they change their jobs as they grow older. During their first days of life they have to clean cells. Then they take over the care of the brood. From the eggs of the queen hatch whitish, helpless creatures that look a bit like fly maggots but change within two to three weeks into winged bees. During the larva stage their foster mothers, the workers, feed them on a very nourishing secretion of their greatly developed salivary glands. Later, when the brood can stand a more robust diet, it is fed on pollen and honey. After having been nursemaids, the workers turn to other occupations. They make their first excursions from the hive, they sweep the hive out, throw out refuse, produce wax, and build new combs, or they do guard duty at the entrance of the hive. After about three weeks the worker bee starts on her final job and goes foraging for pollen and nectar and carries it back to the hive. This the worker does until the end of its life, which in spring and summer, the time of most active collecting, rarely lasts longer than four to five weeks.

In spring the raising of the brood leads to a quick increase in the number of bees within the hive but not immediately to the founding of new bee colonies. This happens when the bees begin to swarm. A new colony needs a new queen. The workers see to it that there is one available. Whether an egg develops into a worker or a queen bee is left to the discretion of the foster mothers. When a larva is brought up in a specially large cell, on large quantities of special food, then we get a fully developed female, the queen. Since she lives for several years while a worker lives for a few weeks only, one has called the queen's food "royal jelly" and has attributed to it health-giving and life-prolonging properties. Royal jelly became the object of a flourishing trade, but whether persons eating it lived longer than their destined span of life may be doubted. But who would not be prepared to invest in so great a hope!

When bees swarm the old queen and half the population of the colony leave the hive and the young successor to the throne takes over. The swarming bees collect in a cluster around the queen, usually not far from the hive. At this moment the alert beekeeper can collect the swarm without much trouble, but he has to be quick. After a few hours the cluster of bees dissolves again and, guided by their scouts, they often fly far away to seek a suitable new home.

This short sketch of life in a bee colony shows how many thousands of its members co-operate in a strictly orderly manner. But the most beautiful example of the high level of organization of their instinctive behavior is what one might call their "language."

How bees communicate with one another

To any observant beekeeper it is soon quite obvious that there must exist some kind of communication in the beehive. For example, he may find that a pot of honey stands out of doors unnoticed for days. But as soon as only one single bee has discovered it, dozens and hundreds of its companions hurry to the vessel to partake of its abundance. They must have told each other at home! How this is done one cannot find out in an ordinary hive. An observation hive has to be built, into which one can look through glass windows. Every returning bee has only to be marked by a dot of color on its back to remain recognizable among the milling crowd on the combs.

Let us put up a glass dish with honey near the observation hive and wait until it is found by a bee. We mark the discoverer of the source of food with a colored dot and watch it after its return to the hive. There it walks up the comb, sits for a while, and hands over its sweet cargo to a fellow worker. Now begins a spectacle so charming that words fail to communicate it. The bee begins a round dance on the comb. It walks in a full circle with quick, small steps, once clockwise, then counterclockwise in quick alternation. This is repeated several times at other places on the comb, and eventually the dancer hastily dashes to the entrance in order to return to the feeding place.

The dance creates great excitement among the surrounding bees. They follow behind and try to touch the abdomen of the dancer with their feelers until they too finally dash to the exit and leave the hive. Soon the first newcomers appear at the dish with honey. It is obvious that it was the dance that raised the alarm. But how did the newcomers find the feeding place?

To bees, glass dishes are unnatural containers. If we replace them in our experiment by flowers, the natural drinking cups of bees, we are in for a surprise.

We feed a few marked bees on a small bunch of cyclamen. In order to sustain a rich supply we put drops of sugar water into the flowers. In the vicinity we place somewhere in the field a bunch of cyclamen and a bunch of other flowers, say phlox. The newcomers, which in their search for food happen to come across the bunch of cyclamen, obstinately search for honey, paying no attention whatever to the phlox. If we now offer the sugar water on the phlox instead of on the cyclamen at the feeding place, the same bees go on collecting it and dance in the hive as before. But at our observation post in the meadow, the picture changes after a few minutes. The approaching bees now search for the phlox only. Even in the neighboring gardens we now see them eagerly visiting the flowers of the phlox. This is a strange sight to a specialist in the study of flowers; he knows that bees never visit the phlox because its honey sits too deep down in the flower and can be reached only by butterflies, with their much longer sucking tube.

From this experiment we conclude that the dancing bees tell

each other not only about the presence of food but also about the kind of flower in which it can be found. To do all this they need not utter one single word. When they collect the sugary fluid from a flower, some of the flower's scent adheres to their hairy coat, and when the fellow workers on the comb walk behind the dancer, touching its abdomen with their feelers, they notice the scent and learn what they have to look for. This is not just a guess but can be demonstrated convincingly. If artificial paper flowers are used as containers for the sugar water and if they are scented with the oil of orange, all foragers from our hive suddenly seek nothing but this scent, although they have never come across it before in their lives.

So far our dancer has used only the simplest means to tell its mates of the existence of a source of food and about the kind of flower that offers it. But other information is given besides. The scout can tell how far its mates will have to fly if the pasture is far away, and which direction they have to take in order to get to the place of successful foraging. The distance is communicated by the rhythm and speed of the dancing.

More wonderful still is the way in which the direction of the food supply is indicated. It is given by the direction of the straight runs during the wagging dance. Since the combs are hung up vertically, whereas the line of flight lies in a horizontal plane, the straight run of the dance cannot directly point in the direction of the food source. Instead, the bee uses the sun as compass reference in its communication. If the food is found somewhere between the hive and the position of the sun, the straight run is directed upward, while a downward run indicates the opposite location of the food. An upward straight run of 60 degrees to the left from the vertical indicates that the food is 60 degrees to the left from the sun's position, and so on (Fig. 61). These instructions are understood and obeyed by the foragers with remarkable consistency. You will ask: What happens when the sky is overcast? It is surprising, but beyond doubt, that even then the bees know the position of the sun in the sky.

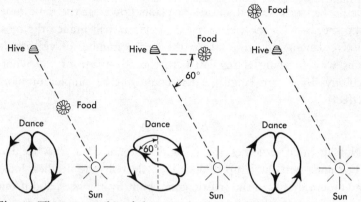

Fig. 61. Three examples of the way the dance of the bees indicates the direction of food. (The sun is southeast of the hive.)

In this way of finding a direction they are much more efficient than most men.

Now to another important point: bees dance only when they find an abundance of food, such as is offered in a glass dish or by flowers rich in nectar. When the supply of nectar dwindles the scouts go on collecting until nothing is left, but they stop dancing and no new helpers are recruited.

This behavior makes sense when we think of the natural course of events. When a species of plant begins to bloom the nectar accumulates in the flowers. When a scout bee discovers this newly laid table it fills its crop effortlessly; its inborn instinct tells it that a dance is called for. The bees in the hive understand the meaning of the dance and away they fly in search of the scent brought in by the scout: away in a search near the hive when invited by a round dance, and farther afield in the direction indicated in response to a wagging dance.

Later, when the foragers are so numerous that the community can easily collect nectar from all flowers within the whole range of flight, the supply eventually begins to dwindle. The dancing ceases and a number of foraging bees, just sufficient for the job, carry on with no further reinforcement from the hive. If, as generally happens, several species of plants flower at the same time, the species with the richest yield of nectar will induce the most active dancing and thus attract the largest number of visitors. In this way foraging is regulated, to the advantage of the whole colony, by a sign language that could not be simpler or more effective.

The ant's nest

Not all insect colonies are as rigidly organized as the beehive. The ant's nest, built up of loosely collected pine needles, looks much untidier than the artistic comb built by the bees, and if one watches the hustle and bustle of the small inhabitants one cannot help feeling that none of them really quite know what they are up to. But this is a superficial impression only, and communal life among ants is quite similar to that of bees.

Here too we find three different castes: the queens, which are fully developed females; the workers, females with underdeveloped ovaries, taking care of the brood, foraging, and defending the nest; and at times there are also males. The males and the virgin queens have wings. On a warm summer day they leave their nest in swarms for their wedding flight up to the bright warm sunlight. The males perish soon afterward; the fertilized queens lose their wings and try to start a new nest, or return and find room in the old one. In this case the "Majesties" get on very well together; in fact, many dozens of queens may be found in one colony, all peacefully occupied with their one task, the laying of eggs. The workers are always wingless. Thus ants are much more bound to the soil and have a way of life quite different from bees.

The great variety in nests and habits makes them especially attractive to the observer. All over the earth we find numerous species and genera of ants, each with its own way of life. Probably best known are the great wood ants, the builders of the largest nests. Other species build smaller heaplike nests in fields and meadows. The brood chambers in the nest are found above and below ground. There the brood is tended: from the eggs emerge little maggotlike larvae. After a period of growth each changes into an ant within a cocoon spun by itself. To the bird lover these pupae are known as "ants' eggs."

Weavers, gardeners, and tyrants

Some species build very unobtrusive nests completely underground or under stones. Many tropical species prefer to build in tree trunks or branches, very likely as a protection against flooding. In Ceylon ants' nests are found in trees: they are made from leaves sewn together with spun-silk threads. This is rather curious as grown-up ants have no spinning glands. Only larvae have them to spin their cocoon with, but they lie as helpless little maggots in their nurseries and cannot crawl around in the leaves to build. The solution of this puzzle sounds like a tall story, but it is founded on the observations of most conscientious biologists. Nests were torn apart to see what would happen. Immediately a crowd of ants came marching along, arranged themselves along the tear, and pulled the rims together. Then from the depth of the nest came some others that held grown larvae in their jaws, squeezing them and making them spin threads of silk. Then they applied the larvae's heads first to one and then to the other side of the tear, thus producing a silk weft, by using the living

Fig. 62. Ants "sewing" a leaf together. Notice the use of the larvae (held in the jaws of the two ants above).

larvae as distaff and shuttle at one and the same time (Fig. 62).

While bees live exclusively on honey and pollen, ants have much more varied menus. They like sweet things, but they also roam about killing caterpillars and other pests in great quantities. This makes them useful to forestry but this is true only of European forest ants. South American planters think quite differently of their leaf-cutting ants. They live in extensive underground nests, climb up the trees, and cut up leaves with their sharp jaws. Processions of them can be seen carrying into their nests these bits of leaves, like umbrellas over their heads. This seems to be a harmless enough activity. The trouble is that they prefer the leaves of crop plants, and as enormous numbers of them collect astonishing quantities of leaves, they have been known to destroy whole plantations. In Brazil wide stretches of fertile land cannot be cultivated because no one knows how to keep these destructive ants at bay. They are the only pure vegetarians among ants, but they do not eat the collected leaves. They cut them into small pieces and make a spongy kind of cake (Fig. 63) of them. This serves as culture medium for a certain type of fungus, which they plant and weed out carefully. They continuously nibble off the sprouting fungus threads and by this mutilation bring about the production of knoblike swellings that are rich in protein and serve as their only food. Since their nests are many feet below

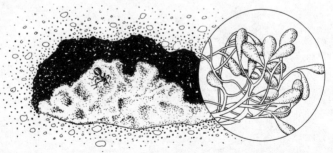

Fig. 63. A mass of fungi (a cake growing on bits of leaves) cultivated as food by ants, and a microscopic view of the fungus (enlarged about 200 times).

the surface of the soil one cannot get at them and Man has to capitulate before these underground fungus growers.

The ever busy ant is a symbol of diligence, but not all ants are made of the same stuff. Some species are slave hunters that raid other ants' nests, steal their pupae, and carry them into their own nest, where the fully developed workers have to labor for their foreign overlords. This can go so far that the slave owners cannot do anything but raid. For everything else they are dependent on their slaves and would starve if they were not fed by them.

Many more things could be said about the colonies of bees and ants, of wasps and termites. But Nature is like a large, wonderful garden, and if we want to see it all we must not linger too long over any single flower bed.

5. *The Balance of Nature*

It is easy to see that the members of an insect society or partners in friendly symbiosis must live in harmony with one another, adjusting to each other's need, lest the peace be disturbed, the same as in a human family. In every friendship worth the name, a harmonious balance needs to be kept to deal with conflicting aims. Even if relations are hostile such a balance can eventually be achieved. The individual encounter between cat and mouse does not end in a compromise, it ends in tragedy. But between all the cats and all the mice in one neighborhood there exists a certain equilibrium of a higher order. When there are many cats the mice are kept in check; remove the cats and the mouse population grows. This is so obvious that it hardly seems worth mentioning. But such obvious facts when overlooked can suddenly pose important problems demanding all our attention.

Small causes, great effects

In the year 1886 California had the most flourishing orange and lemon groves. However, with the importation of plants from Australia a small, inconspicuous parasite was introduced, the citrus scale insect, a relative of the green fly. The new habitat in California suited it well. It multiplied to such an extent that many orange and lemon groves succumbed to its mass attack, and showed reduced yield. For a while all countermeasures taken were unsuccessful. The planters were desperate and began to fell the trees.

The German entomologist Koebele was sent to Australia to find out why the citrus scale did not do any substantial damage in its native country. He found that there it had a natural enemy in a certain species of ladybird that kept its numbers down, but unfortunately this insect had not been introduced into California. It seems that ladybirds the whole world over show a similar taste. The Australian citrus scale, however, could be fought only by the Australian ladybirds. Koebele sent a consignment of them to California, but they did not survive the long journey. Then on his return he himself brought back a hundred living specimens. He let them loose on an orange tree covered with cheesecloth, and they at once attacked the citrus scales. On this rich food supply they multiplied and their numbers rose to ten thousand within a year. These were handed out to the farmers and liberated in their orchards. Within two years the ladybirds had reduced the number of scale insects to such an extent that the trees declared incurable showed new life and a rich yield in fruit.

The citrus scale has never since then regained the upper hand in California. Of course the ladybirds too stopped multiplying at the former rate, because of the dwindling numbers of the citrus scale. The relation between pursuer and victim can be likened to the laws of supply and demand. If a biological equilibrium, such as exists between the California scale insect and the ladybird, is upset, a disturbance of ever-widening consequences may fol-

low. Take our very clear example: without the introduction of
ladybirds the citrus scales would have multiplied at such a rate
that the complete destruction of orange and lemon groves would
have deprived them of their livelihood and both scale insects and
orange trees would have finally perished together in the flames lit
by desperate planters.

It is very rare that an animal like the citrus scale has one
outstanding enemy only. Deer, for instance, are hunted by Man,
foxes, martens, and birds of prey alike, and when it was thought
that the number of deer could be increased by keeping the other
predators in check by destroying them, the very opposite happened. It was realized too late that chiefly sick and weak deer
were eliminated by these predators, much to the advantage of
the healthy ones, and that the whole deer population may suffer
if such wholesome natural selection is prevented. The relationships in a natural habitat are so delicately poised that interference
with this balance may have unpredictable consequences.

The living community of the forest

Animals and plants of any natural habitat are interdependent.
Under normal circumstances they form well-balanced living
communities. But it is not only the interrelations of organisms
that are important; so too is their relation to environmental factors, such as light, air, and soil — a relationship that can work
both ways. When the Romans felled forests of age-old trees along
the Adriatic coast to build ships, and when later the Venetians,
doing the same, used up what was left of the forest, the rains
washed away the soil from the unprotected ground and the heat
of the sun baked it dry. Where before, in the shadow and the
moist air of the grown woods, the young trees found suitable
conditions, none of them could now grow up successfully. Such
was one of the causes of the barren Dust Bowl of the American
West. Only very recently have we succeeded in redressing the
balance by reforestation, with much cost in time and money.
Here cause and effect were so obvious that the lesson was easily
learned.

However, another mistake was made, because the laws of a living forest community were incompletely understood. A natural wood is a mixed wood, where various types of trees grow together at random. A rich bacterial population slowly changes the fallen foliage into humus and a new food supply for the trees. In the tops of the trees all kinds of caterpillars and other parasites find plenty of food while their numbers are kept in check by birds, by hordes of wasps and other enemies. The more varied the kinds of trees, the more varied the living conditions for the occupants; the more varied the insects and other animals that live there, the more varied the species of birds that find suitable breeding places and living conditions. All things are interlocked, and an occasional localized disturbance need not upset the community as a whole, because in spite of individual fights and dispossessions the self-supporting whole remains in balance.

Now Man comes onto the scene and finds that some trees grow more quickly than others and are therefore more profitable to him. So he starts to fell the mixed woods and to grow just one kind of tree. For a time everything seems all right, and the new method appears to pay. But the uniform needs of the planted trees lead to a one-sided usage of the soil and to its deterioration. A large proportion of the natural population of a mixed wood, birds and insects, do not find pine woods or fir woods to their liking and move away or perish. The parasites and their natural enemies are reduced to a few species, but their individual number increases. And if, as can easily happen, the external conditions favor the parasites for a number of years in succession, they multiply unchecked and damage of unforeseen extent can follow. Nowadays we are well aware of the advantages of a mixed wood and the uniform pine and fir forests are on the way out, much to the relief of the naturalist, who never liked that monotonous roof above his head.

Primeval forests in all their beauty will never again grow in our part of the world. Yet within the framework of progressive civilization there is scope for a sensible conservation of Nature.

As our examples concerning hunting and reforestation show, this is not a matter of sentiment only but of common sense. Slowly we come to understand that Man cannot afford to forget that he himself is but one link in Nature.

IV

REPRODUCTION

When I was a youngster my room was like a small zoo and my activities were encouraged and fostered by understanding parents. From sea anemones and all kinds of creepy-crawlies, right up to a parrot and a mongoose, all families of the animal kingdom were represented and offered opportunities for all kinds of observation. I remember one incident. I got a pair of canaries for a present. Soon there were eggs and I was filled with expectation; alas, nothing ever hatched from them. Without really knowing why, I separated the two birds. Soon there were eggs in both cages. At the time this matter was not at all clear to me, but the rest of the household were rather amused and I was told the egg of a bird can develop only if a male has been present.

It is perfectly true for birds, mammals, and other animals that reproduction is possible only when males and females meet, but it is far from being a law of Nature. Among lower animals reproduction generally is not linked up with sex at all. One speaks of asexual reproduction in such a case, and we want to discuss this first.

174

1. *Asexual Reproduction*

In the first chapter of this book we talked about amoebae and other microscopically small organisms, which do not know old age or natural death. If they lived in the Garden of Eden, they would not have to propagate. Life being what it is, losses within their ranks are unavoidable and they have to make up for them. This they do. As a rule they multiply by simple cell division. But on closer inspection we find that this is not quite so simple as it looks at first glance.

Nuclear and cell division

The dividing cell passes on its substance and material to the newly formed daughter cells. The protoplasm of a cell is as easily halved as a dinner roll by a constriction running through the middle of the cell. But the important activity in the division of one cell takes place in the nucleus. The nucleus contains chromosomes, bodies that take stains (from the Greek *chroma*, color, *soma*, body). If all the chromosomes did was to take stain we might disregard them. But it is through the chromosomes that both daughter nuclei can receive the same set of hereditary factors, which are located in the micro-structure of the chromosomes.

If heirlooms are to be distributed fairly between two heirs, they have to be divided in lots. It would not do, of course, to cut each article in half — that would ruin them. But in the living cell this is exactly what happens. In each chromosome are certain units of genetic material (Fig. 64). These units, made up of a substance called DNA (which we shall deal with later), can produce exact copies of themselves. This doubling of the genetic material finally results in a longitudinal split within the chromosome. Thus we have in each fully grown chromosome two identical halves. They separate and one of each is allocated to each of the daughter nuclei. Thus it is one of the characteristics of living cell material to be able to grow and to replicate itself. In this way a truly just division becomes possible.

This is the reason for the remarkable chain of events that we observe during the duplication and division of the chromosomes. A few more details are worth noting: as a rule we find in the protoplasm besides the nucleus a tiny granule, known as the centriole (Fig. 64a). This too divides and its two halves move apart. The nuclear membrane disappears after the chromosomes have become visible (Fig. 64b). After this the chromosomes assemble in the middle of the cell between the two centrosomes (Fig. 64c). They are attached to fine, threadlike structures originating at the centrosomes. When these structures shorten, the daughter halves of the chromosomes are pulled apart. After that the two groups of daughter chromosomes disappear from view, while new nuclear

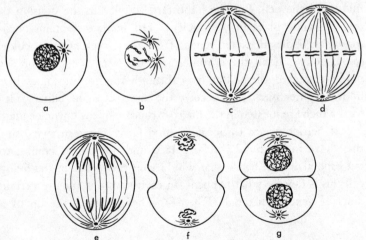

Fig. 64. Division of the nucleus and of the cell. The cell has four chromosomes.

walls are formed around them. The division of the nucleus is followed by a division of the protoplasm.

Centrosomes are not always present. They occur more regularly in animal cells than in plant cells, where in fact they may be present but may be too small to be visible. In one and the same species the number of chromosomes appearing during cell division is always constant, while different species of animals and plants have different amounts. The sequence of events during cell division in animals and plants, especially the duplication of the chromosomes, is the same in simple and in highly complex organisms.

In unicellular animals and plants nuclear and cell division leads to multiplication and reproduction of individuals. In multicellular plants and animals the manner of cell division is the same, but the newly formed daughter cells stay together. Whether we study the cells in the root of a plant or those of an earthworm or of a human being, every time we see chromosomes appear, split lengthwise, and hand on an equal measure of their contents to the nuclei of the daughter cells.

Regeneration

In multicellular organisms cell division is responsible for the growth of the individual and not for its reproduction. Reproduction by cell division does occur in some lower multicellular animals, which can replace large parts of their body by a process known as regeneration.

Some achieve incredible feats. In ponds and brooks live small worms, about a quarter of an inch in length, that are known as flatworms or planarians. Their two eyes are two black dots in front (Fig. 65), and their mouth is situated in the middle of their belly and opens into a large voracious gullet. If one cuts such a worm into three parts one by no means kills it. Within a few weeks the severed head will produce a new hind part complete with mouth and gullet, the severed tail-bit a new front part with mouth and eyes, and the middle piece will produce a

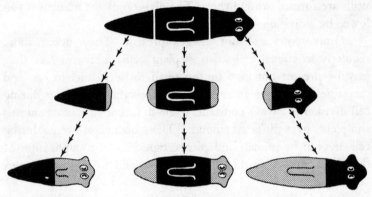

Fig. 65. Regeneration of a planarian from portions of the head, body, and tail region.

new head and tail. This regeneration of lost parts of the body is possible in lower animals because not all of their cells have definite tasks allocated to them. Besides muscle cells, sensory and nerve and gland cells, many nonspecialized cells exist. They are endowed with all the original potentialities. They migrate to the place of an injury and regenerate the lost parts. In this case mutilation of the body can lead to a multiplication of individuals.

Multiplication by division and budding

But not only brute force stimulates these animals into showing what they can do. Some break in halves of their own accord and replace the missing bits. As a rule the new head and tail have already begun to grow before the animal breaks in two, with the result that the two halves can grow all the faster into two complete animals. Nevertheless this halving of the individual is bound to be upsetting to its daily routine. This is probably the reason why in other species this process is somewhat modified and replaced by a more progressive method known as budding. This is the case with Hydra, the fresh-water polyp that we have already described. It, too, can halve itself and replace the missing parts with the help of nonspecialized cells. As a rule these cells collect at one side of the body and produce a new polyp, which sprouts like a bud and eventually drops off, without interrupting the food intake or any of the other routines of the daily life of the mother animal.

In plants this kind of multiplication is even more common than in animals. Unicellular plants divide like unicellular animals. Multicellular plants can be cut up and will produce complete plants from parts. The gardener makes use of this by planting parts of shoots and by growing from them many complete new plants. Spontaneous halving of multicellular plants occurs, but budding is more common. Highly organized plants form buds that drop off and grow into new plants. Other plants, such as the strawberry, produce runners (really horizontal shoots), on which new plants develop. In the potato, underground stems produce tubers that carry buds.

In higher animals, including Man, the replacement of lost parts is restricted to the modest achievement of wound healing. We cannot even replace a cut-off finger tip, for all our cells are highly specialized. We lack the "nonspecialists" and with them the sources for regeneration and asexual propagation by division and budding. This is the reason why this type of multiplication does not occur in the highly developed arthropods and vertebrates.

2. Sexual Reproduction

Egg cells and sperm cells

The best known, the most frequent, and, in higher animals, the only way of multiplication is reproduction by eggs. In most animals we find two sexes, males and females. They look outwardly different but this is not necessarily so. Internally they always differ in one respect: the females form eggs. In lower animals eggs appear in different parts of the body, but usually they develop in a glandular organ only, the female gonad or ovary. The corresponding organs in the male are the male gonads or testes. They produce the sperm cells.

As a rule an egg develops only after its nucleus has fused with the nucleus of the sperm of the same species. This fusion is called fertilization and is a necessary step in bisexual reproduction.

Some animals can develop from unfertilized eggs, a process that is known as unisexual reproduction or parthenogenesis.

The importance of eggs in reproduction has been known for a long time. Eggs are often quite big, as in fish and birds, and nobody can possibly overlook them. But what actually happens during fertilization has been known only since 1875. The sperm cells are very small and are visible only under the microscope. They had been discovered long before the role they play in reproduction was understood. As so often in the history of science, it was the study of the simple lower animals that yielded decisive and revealing results, without which it would not have been possible to understand reproduction in highly developed animals and in Man.

Fertilization of the sea urchin's egg

The dark, prickly sea urchins are common animals living in the shallow waters near the shore. It was on those animals that the brothers Oskar and Richard von Hertwig made their classical observations in the seventies of the last century. However far away from the sea a zoological department may be, sea urchin eggs will be used in demonstrations in which students of biology are shown the miracle of fertilization and consequent early development under the microscope. They are the living material best suited for this purpose.

When the time for reproduction comes, the male and female sea urchins cluster together in great numbers and discharge their germ cells into the sea. Egg and sperm are of different size and shape. The egg cells contain a lot of nourishing material in their protoplasm in the shape of yolk grains. This serves as food for the developing embryo until it can gather food by itself. This ballast makes the egg cell relatively large and immobile. It is left to the male sperm to seek out the egg cell and fuse with it. The sperm cell is well equipped for this task. Its cell nucleus is covered by an extremely fine membrane only, and has hardly any protoplasm attached to it. The sperm nucleus is compact and small because all unnecessary water has been given off. In fact, it forms the head of the sperm. The sperm is propelled by the whiplash movement of a tail.

The eggs secrete a substance into the water that attracts the sperm cells and makes them move more actively. This substance has been extracted and studied. In its purified state it is a red pigment, chemically related to naphthalin. If one puts a trace of it into a glass tube and places this in sea water containing sperm cells of sea urchins, they swim right into the glass tube in the way they usually approach sea urchins' eggs. The effective radius of the substance is great. If two pounds of it were dissolved in a round basin of water, two thirds of a mile in diameter and filled to a depth of six to nine feet, one drop of this solution would still attract sperm cells. The eggs of a million sea urchins would yield two pounds of the pure sperm-attracting substance.

The efficiency and specificity of this substance remind one of hormones, with which it has other properties in common. So far such substances have been studied in very few animals and plants, but we can expect to find them everywhere. The egg sends them as messengers to show the sperms the way.

When a sperm meets an egg its head penetrates into the egg protoplasm (Fig. 66a). The tail is thrown off and perishes (b). It has done its duty. The head, which is the nucleus of the sperm, imbibes water and swells until it has regained the shape of a normal nucleus. It then approaches the nucleus of the egg (c) and fuses with it (d). Thus the chromosomes of both nuclei are

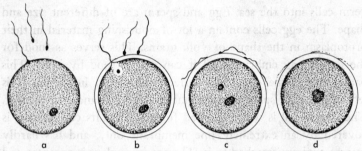

Fig. 66. Fertilization of the egg of the sea urchin.

united. We have already mentioned that the chromosomes are the carriers of the hereditary factors that shape the specific and individual characters of a living organism.

The fusion of the nuclei of egg and sperm cell brings together the hereditary factors of the male and female parents, and this is the most important thing in fertilization. It explains why the progeny shows both paternal and maternal characteristics.

Immediately after the entrance of the sperm cell a membrane is laid down on the surface of the egg that prevents other sperm cells from entering. This is important because the entrance of several sperms might lead to malformations. As soon as the egg is fertilized the development of the embryo begins. Unfertilized eggs perish after a short time.

The knowledge gained from the sea urchin is of fundamental importance. The sequence of events is the same in lower and higher animals and even in plants. The essential features of reproduction are the same in the whole of the animal and plant kingdoms. Differences exist, but they are not fundamental.

External and internal fertilization

Most fish aggregate like sea urchins during the breeding season and discharge their germ cells into the water. It is not surprising that not all sperms reach their destination, and that some eggs remain unfertilized. In the fish hatcheries artificial fertilization is carried out. Females that are ready to spawn are caught, the eggs are squeezed from their bodies into a bowl, and neither fish nor eggs are damaged. Males are similarly dealt with, and the mixing of eggs and sperms yields a high percentage of fertilized eggs.

A similar safeguard for successful fertilization is found in Nature. The eggs and larvae of frogs develop in water. The adult frogs therefore collect in the breeding season in ponds and water puddles, where their love serenades can be heard near and far. Each male frog seeks out a female, very often several days before she lays her eggs. The female is clasped firmly between the front legs of the male (Fig. 67). This may go on for days, and the male is equipped for it by growing at this time thick calluses on its thumbs, which it presses into the skin of the female. When at last the eggs are shed, the male assists by the pressure of its firm embrace and pours its sperm over the eggs as soon as they appear, thus assuring perfect fertilization.

This is what is called external fertilization. The fusion of the germ cells is, however, assured to a higher degree by internal fertilization, or copulation. This involves the transfer of sperm into the passage by which the eggs leave the body of the female. Copulation is known in some aquatic animals, but in terrestrial

Fig. 67. Mating in the frog. Left, the forearm and fingers of the female; right, those of the male, with calluses on one of the fingers.

animals it becomes essential, because the sperm cells with their lashing tails can propel themselves only in a fluid, not in air.

For the transfer of the sperms some most extraordinary devices and instincts are employed. Thus the males of spiders have at the end joint of one of their feelerlike head appendages a hollow, bottle-shaped structure with an opening at its slender tip. Before mating they weave a silken web onto which they deposit a droplet of semen, in which the sperms swim about. This droplet they tuck in the little bottle, the neck of which they then insert into the female genital opening, injecting its contents into it.

These strange forms of sperm intake are the exception. As a rule the male genital opening is simply pressed against or introduced into the female opening. This is the case in most insects and in terrestrial vertebrates. The males very often have special copulatory organs in order to ensure a reliable transfer of semen. After all it is on this safe transfer that the propagation and with it the survival of the species depend. The equipment with sexual organs alone would, however, be of no avail if it were not for the powerful urge that attracts the sexes to each other. These are the most formidable instincts that Nature has implanted into the animal world. Only Man is able to master them by the power of his will and the taming force of love.

3. *Parthenogenesis and Hermaphroditism; Scope and Significance of Bisexual Reproduction*

A world of spinsters. Are males dispensable?

We have already mentioned that some animals have unisexual reproduction or parthenogenesis, which means virgin birth. The colonies of plant lice, so disliked by us on the young shoots of

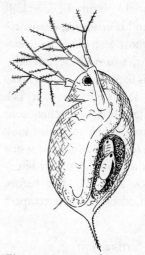

Fig. 68. A water flea with young in her brood sac.

roses and other plants, are mostly communities of females only. Their eggs develop without a single male making its appearance. The same happens in water fleas, which live in ponds and lakes. They are not parasites like the fleas on land, but are harmless little crustaceans, whose only likeness to fleas is their power of leaping. They appear in masses and yet for months on end one cannot find a single male among them. The eggs move from the ovary into a sort of brood pouch at the back of the animal (Fig. 68). There, without being fertilized, they develop into females which in their turn reproduce parthenogenetically again. Some butterflies, worms, and other animals reproduce in a similar fashion.

Why are there males at all if some animals can manage without them? As long as plant lice and water fleas occur in the female form only, every single individual produces eggs, and the multiplication rate is quite phenomenal. For the continuation of the species and its dispersal isn't this arrangement more advantageous than if half of the progeny were males, which cannot produce eggs but only fertilize them? Could the answer be that in some animals unfertilized eggs cannot develop at all? As we shall see this answer is not quite satisfactory, for two reasons.

First, biologists have indeed succeeded by artificial means in making eggs develop parthenogenetically in those animals in which development is normally preceded by fertilization. For instance, sea urchin eggs can be stimulated to develop into full-sized adults by adding certain chemicals to the water, instead of sperm. By pricking the eggs of frogs with a very fine needle, development can be initiated, and they grow into tadpoles and mature frogs without fertilizing. Why should Nature be less skillful in doing what even Man can achieve in his crude experiments?

Secondly, it is known that among most animals which normally reproduce parthenogenetically females turn up occasionally whose eggs have to be fertilized. When this happens males also make their appearance. Thus during spring and summer plant lice reproduce parthenogenetically, while in autumn they change to bisexual reproduction. For weeks and months one may also look in vain for male water fleas. But there too comes a time when they make their appearance. All this seems to indicate that bisexual reproduction is of great importance after all. But before inquiring any further into its real significance, let us attempt a more thorough investigation.

Hermaphroditism and cross-fertilization

At fairs all kinds of oddities of Nature are shown, and among them sometimes a human hermaphrodite. It is very rare that male and female germ glands occur in the same human individual. Such a person is a freak and cannot produce any progeny.

But among lower animals hermaphroditism is very common. Every earthworm that we dig up in our garden, every slug that crosses our path, is both male and female at the same time, and so are many other animals. Among the vertebrates only some fish are hermaphrodites.

One might be led to think that in hermaphrodites fertilization is most effectively ensured, provided the sperm cells of the animal can reach its own eggs. Strangely enough, this does not happen very often. In spring one can see pairs of snails embracing each other, but each partner acts at the same time as male and as female. At the climax of their courtship the partners drive with great force a sharply pointed calcareous dart into each other's flesh. The exchange of this love token seems to be necessary for the stimulation of the cold-blooded lovers, because its injection is followed by the exchange of semen into each other's female genital ducts. In this manner the eggs are cross-fertilized and self-fertilization is avoided. In other animals the male and female germ cells of the individual ripen at different times, the animal being first a male and later a female or vice versa, and again self-fertilization is prevented.

Here, as in unisexual and bisexual reproduction, the question arises why Nature chooses such seemingly complicated methods rather than simple ones. We have to keep in mind that it is not fertilization as such, but cross-fertilization, the exchange of germ cells between individuals, that is the aim of sexual reproduction.

The significance of cross-fertilization

The great importance of cross-fertilization is underlined by the fact that it is not just the special feature in this or that animal or species but is apparent throughout the whole of the animal and plant kingdoms with the regularity of a natural law. Let us consider a few examples.

We have just learned that animals that can reproduce unisexually turn to bisexual reproduction at regular intervals. The fresh-water polyp, which can multiply within a short time by budding, is not content with this method alone but occasionally produces eggs and sperms. From the fertilized eggs new polyps develop and asexual reproduction goes on again for a number of generations. The marine relatives of these polyps stay together and soon form widely branching colonies. But from time to time certain buds turn into individuals of a shape so different that one is tempted to believe them to be a different species. They are

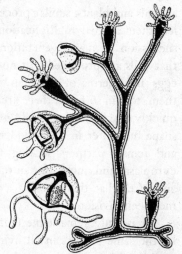

Fig. 69. Colony of polyps forming medusae. In the foreground, left, a detached medusa.

medusae. Medusae are of two kinds, male and female. They swim away from the colony and reproduce bisexually (Fig. 69). From their fertilized eggs new polyps develop. These are the founders of new colonies. Here the change in the method of reproduction goes hand in hand with a change in the form of the animal. In plants too, where asexual reproduction is still more common than in animals, bisexual reproduction intervenes. In flowering plants the female germ cells are contained in the ovules, the male germ cells in the pollen grains. When the flowers are pollinated by the deposition of pollen of the same species on the stigma, the male germ cells, contained within the growing pollen tube, reach and fertilize the egg cell through the style. In hermaphrodite flowers that produce pistil and stamens, that is female and male germ cells at the same time, we find, as often and persistent as in the animal kingdom, the most intricate mechanism designed to prevent self-fertilization and to guarantee cross-fertilization. Egg cells occur in lower plants too, for example in fungi, mosses, and ferns.

Even where one would expect them least, in the unicellular animals and plants, similar processes occur. We have already studied their asexual multiplication by cell division leading to the succession of many generations. Then suddenly it can happen that cells are produced that no longer divide but behave like the eggs and sperms of the many-celled organisms. They resemble them, too, in so far as there are larger, inert female and small and quickly moving male cells. In other cases they are alike in size and shape but differ in their hidden features. They function as male and female partners seeking to unite with each other. Their cytoplasm and nuclei fuse in the same way as in eggs and sperms, and after that there begins a new period of multiplication by mere cell division.

It is truly remarkable that fertilization should, in principle, be the same process in all living organisms, including Man. We shall be able to appreciate the full implications of this when we discuss the mechanism of heredity.

What, then, is the significance of fertilization? Science cannot give a complete answer to this. In higher animals there is no reproduction without fertilization, and development begins immediately after the fusion of egg and sperm. This made people think that fertilization plays the part of the necessary stimulus for embryonic development. However, this cannot be the whole story. For instance, parthenogenesis, in itself a quite efficient mode of reproduction that dispenses entirely with the stimulating effect of the sperm cells, is periodically replaced by bisexual reproduction. Observations on unicellular creatures also clash with this idea. After prolific multiplication by cell division fertilization suddenly intervenes, only to become dormant again for a long period of time, which is the very opposite to a stimulation of developmental processes. Besides, in unicellular organisms fertilization always leads to a reduction in the number of individuals because two of them fuse to make one.

The safeguards to secure cross-fertilization in animals and plants point to an explanation. Nature's aim is not just the fusion of germ cells but the fusion of germ cells of two different individuals. Their hereditary make-up is never quite the same. As a consequence of their previous history their hereditary factors show many kinds of small differences. In cross-fertilization these differences have a chance to appear in all manner of combinations, and this makes for variation in animals and plants. Variability is very useful and important. Over long periods of time external living conditions are bound to change, and as we shall learn, the variability in hereditary characters is the main premise for successful adaptation of living organisms to environmental change. Here is the key to the prevalence of cross-fertilization and of sexual reproduction.

4. *Courtship, Engagement, and Marriage in the Animal Kingdom*

In multicellular organisms chemical attraction guides the sperms to the egg. Before this can happen the parents have to find each other. In hermaphrodites both partners are equally engaged in this. Where the sexes are separate the individual as a whole behaves like its germ cells: the males are more lively and take the initiative, the females are passive. The seeking, finding, and courting between the partners is here an affair of the senses. Of these all may be engaged but foremost among them are smell, hearing, and vision. Hence the differences in external appearance and behavior of male and female; these are known as secondary sex characteristics.

Two things are important. After male and female have found each other they must not confront each other with indifference, but they have to get excited. Eventually germ cells are made to meet and fuse for the sake of the survival of the species. Let us choose a few examples from the many offered by Nature.

We have already described how the female moth attracts the male from afar by scent. By such simple means it achieves a success that could fill the most seductive woman with envy. One single female of a rare species of moth attracted, within seven hours, one hundred and twenty-seven males. They were, however, thwarted and all found their end in the insect cabinet of a collector who was shrewd enough to use the female as a bait. Many mammals have scent glands in different parts of their body and scent plays an important part in courtship. Scent glands play a role during courtship of aquatic animals.

In spring one can observe male newts following the females about for hours. Occasionally a male will overtake a female with quick steps, jump in front of it, and squat down, wagging its tail and fanning water toward It. This is saturated with the scent from glands on the male's belly, a scent that makes the female willing to accept the courting male.

The production of sounds and noises during courtship is widely used in higher animals. We have already learned how crickets and grasshoppers call to each other. Among the vertebrates the song of birds and the less melodious choruses of frogs are well-known examples of how the males vociferously inform the females of their presence.

Most familiar to us are those sex characteristics that catch the eye. Males, and more rarely females, are adorned with colored patterns and ornaments of body and appendages. To animals with good eyesight these are as sure a means of recognition as the specific scent to animals with a well-developed sense of smell. Their significance to the female can be gauged from the emphatic manner with which they are displayed by the male.

Thus the male of the fiddler crab has one claw tremendously enlarged and vividly colored. If at the mating season a male meets a female it raises itself high up and agitates this big shining claw with tremendous speed, beckoning the female to join it. Similar signals are used by spiders.

The male newt, apart from its scented message, displays its splendidly colored skin and stately crest. A colored wedding dress at spawning time, quite common in fish, is also exhibited for the benefit of the female.

Birds show the whole gamut of courtship, from the simple bowing of a modestly clad male to the most dazzling display of colors in others, to which may be added acrobatic feats, fierce sexual fighting, and ritual dancing, sometimes in elaborately constructed love bowers.

The penguins, those amusing creatures of the antarctic region, are rather phlegmatic lovers. The female digs a hollow for a nest and then waits patiently until a male arrives. The moment it presents itself the marriage is consummated. The male then begins to collect stones, which the female uses to build a wall around its nest.

In other birds the choice of a mate is preceded by a long courtship. The peacock is well known for fanning its tail in a performance that is even surpassed by other birds, such as the argus pheasant. This displays its marvelous wings before the female and at the same time does not miss the opportunity to peep through a gap in its plumage to watch the impression it is making. There is also the bird of paradise, which affords its mate a special surprise by suddenly displaying its splendid plumage in a most unusual posture, hanging head downward from a bough.

The male of other birds of paradise builds a special nest, in a clearing of the wood, a so-called bower of branches, into which it carries colored snail shells, stones, feathers, bones, even fresh flowers. When the bower is finished the female enters and the male twists its body in strange contortions, at the same time producing extraordinary sounds. Occasionally it stops to pick up a feather or a flower with its beak as if to point out its treasures to the female. This courtship is performed with such ceremonial grace and apparent aesthetic sense that it looks almost human.

Other birds are much less poetically minded: they seem to practice cupboard love only. The male of some terns first catches a fish and then mates the female, which accepts the prey from his beak. The tropical cuckoo courts its female by approaching it with a locust in its beak, which, however, it parts with only after the female has obliged. Anyway, this is a more pleasant wedding feast than that practiced by certain locusts and spiders, among whom it is the habit of the female to devour the male after copulation.

Male hotspurs among birds fight each other fiercely during courtship. This has its mammalian counterpart in the fighting of male red deer in rut. Their antlers turn from ornaments into most dangerous weapons. And yet this is just another form of courting the female. The fight may be a harmless spectacle, but it may also lead to a bloody climax.

Marriage on a permanent basis

A lot of fighting takes place when polygamous males have to defend their harems against rivals. But monogamy is the rule wherever permanent relations are established. In lower animals we rarely find anything resembling marriage, because the sexes do not live together for any length of time. A most remarkable exception is the dung beetle, whose female chooses its partner from among several eligible applicants. After that the pair lives together for the whole of the summer. In insects constancy is ruled out by the fact that the adults live for only a few weeks or months.

The most perfect marriage relations are found among birds. Some, like our ducks, pair off in autumn, though they do not reproduce until the following spring. They are thus engaged for quite a long while. Others go their own way during the winter but reunite during the breeding season. Wild geese, ravens, and probably some of the owls lead a regular married life and stay together summer and winter. But there are exceptions. The male and female of the great woodpecker continuously bicker with one another and behave as if it were extremely distasteful to both of them that two are needed to produce offspring and to care for the young. In mammals other than Man permanent relations are rare except in apes.

Fig. 70. Female of a deep-sea fish (Edryolychnus) with three males implanted in her.

Honesty compels the admission that not always are beauty and strength the attributes of males. In certain deep-sea fish young males attach themselves to the female and grow onto its body (Fig. 70). Their sense organs stop functioning and they lose their teeth. Their blood vessels fuse with those of the female and henceforth they live on its blood only. They have become parasitic appendages. Their body remains dwarfed lest they weaken the female. Only their sperm-producing organs, their gonads, are well developed. Quite possibly it is difficult for a male to find its mate in the vastness of the deep sea. To grow onto one's female, so as never to lose it, is not a bad solution. It is left to future explorers of the deep sea to find out how its other inhabitants find each other and arrange to meet again.

V

DEVELOPMENT

---

1. *From Egg to Chick*

An appetizing egg, sunny side up, stands on the breakfast table. How quickly it will be eaten! And yet let us be aware that this is not its natural purpose. If Nature could have taken its course, a young chick might have hatched from it, within three weeks of incubation. Let us therefore turn our attention away from breakfast egg to the egg as an animal cell, and a germ cell in particular, destined to produce new life.

In the very center of the egg lies the yellow yolk. This is the true egg cell which is formed in the ovary of the bird. From there it proceeds to the outside world, passing through the oviduct. While the egg descends slowly, the gland cells of the oviduct deposit a nourishing coat of egg white around the yolk. Lastly a thin shell membrane and, as a final protection, a calcareous shell are added.

At first glance the yolk seems to be a uniform substance. Only on closer inspection do we see at a certain place on its surface the egg nucleus surrounded by a small zone of protoplasm. This structure is called the germinal disk (Fig. 71). It contains all the developmental powers. Everything else is merely an

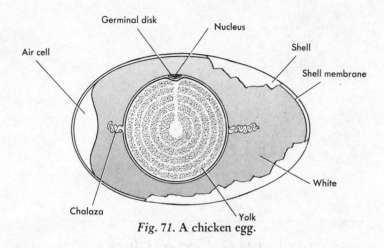

Fig. 71. A chicken egg.

adjunct: numerous yolk globules consisting of protein, fat, and oil droplets, which represent the nourishing substances embedded in the cell. These adjuncts make up the main bulk of the egg and they are what we are after when we eat it. But they are of little use in the study of the process of development; indeed, the presence of such large quantities of yolk in the cell variously influences and obscures the process. Let us therefore leave the chicken egg to the cook and study instead the developmental processes in the tiny lancelet, a fishlike animal with small eggs and little yolk. In spite of all the differences in final shape, animal development follows a common trend. Thus, as with fertilization, the study of lower animals leads us to a better understanding of the functioning of more highly evolved organisms.

A simple example of animal development

A lancelet is a marine animal and is only about two and a half inches long. Despite its unassuming looks, it is a famous animal because its design shows a number of fundamental features that have contributed a lot to our proper understanding of the vertebrate body. Actually the animal is not a fish at all and some of its features remind one rather of a worm; for instance, it has a front end but no head. However, we do not want to go into details of design; all we want to describe is how from one single egg cell an animal's body with all its organs is formed.

The eggs are deposited in the water and fertilized there. We know that during fertilization the nucleus of the sperm fuses with the nucleus of the egg. This is followed by development, and the first thing that happens is a sequence of cell divisions, the nucleus always dividing before the protoplasm. Thus, by repeated division, an egg cell is transformed into a cluster of cells that become smaller with every division, since, at that stage, the embryo does not yet take up any food. It is the material of the original smooth egg cell that cleaves into many cells (Fig. 72 a-d). These eventually arrange themselves in one single layer on the surface of a ball-shaped structure, the inside of which is hollow and filled with fluid (e). This stage in the development is called the blastula stage.

Now this structure begins to take shape. First, at a certain point, the wall of the hollow sphere turns inside until it touches the opposite wall (Fig. 72g). Thus the one-layered blastula turns into a cup-shaped two-layered structure, the gastrula (g, h). At this stage the lancelet consists, as it were, of a skin covering a hollow gut. We say it now consists of two germinal layers. From the inner germinal layer, the endoderm, develop the tissues of the gut and the organs of the alimentary system. The skin is formed of the outer germ layer, known as the ectoderm. The opening of the gastrula cup is called the blastopore, and it turns eventually into the anus of the animal, while its mouth is formed in a different place and breaks through later.

The ectoderm produces not only the skin but other parts of the body such as the brain and the spinal cord.

In the fully grown animal the brain and spinal cord do not strike one as related to the skin, in position or function. But we must realize that they are derived from the skin at a very early developmental stage and then go their own way. It happens like this: the embryo elongates and on its back appear two ridges

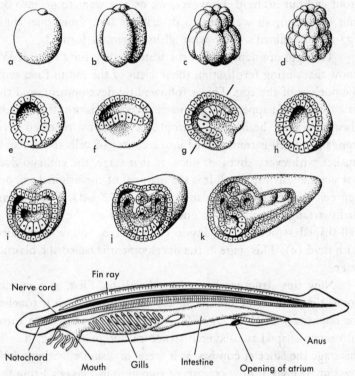

Fig. 72. Amphioxus (the lancelet) and the principal stages of its embryonic development: e-g, longitudinal section; h-k, cross section. (Amphioxus and stages much enlarged. Amphioxus itself is about two and a half inches in length.)

with a groove between them. The material of the groove sinks below the skin and forms a tube that turns into the brain and spinal cord, while the skin grows over it and covers it completely (i, k).

At the same time two lateral outgrowths arise from the original gut wall. They are eventually folded off from it to form a middle germinal layer, which comes to lie between the outer and inner ones. This is called the mesoderm. The third germinal layer does not just form simple tubular structures to the right and left of the gut, but subdivides into a row of sacs (j, k). At the dorsal side of the gut, exactly under the nerve tube, a rod of tissue is pinched off and turns into a long, solid rod of cells.

Thus the cells of the developing embryo arrange themselves into tissue strands, tubes, and sacs, but at this stage one cannot guess their ultimate fate. Eventually the cells take on the shape and arrangement suitable for the performance of their final functions, and the organs, so far only sketchily outlined, begin to develop their final shape.

For example, the tissue strand between gut and nerve tube turns into an elastic supporting rod, corresponding to the spinal column of higher vertebrates; the sacs of the median germinal layer produce the body muscles, connective tissue, and blood vessels; the remaining gut develops into the alimentary tract. We do not need to discuss further details of organization here.

Yolk complicates matters

In the development of the lancelet we can roughly distinguish three main stages, which, however, are not strictly set apart. In the first stage, during the cleavage of the egg, there is a multiplication of cells; in the second stage the cells arrange themselves into germinal layers; and in the third stage the germinal layers form tissues and organs fitted for their future tasks. It is interesting that in all animals the development shows the same fundamental sequence. The existing differences have to do mainly with the varying amount of yolk in the egg cells.

The egg of a newt contains many more yolk globules than the egg of a lancelet. Since they are not regularly distributed but collect predominantly at one pole of the egg, the cleavage cells are of different size, small at one pole where there is little yolk, large at the other pole where most of the yolk is concentrated (Fig. 73 a-d). This yolk influences the formation, shape, and position of the gut cells (e-g). The mesoderm also develops in that restricted space in somewhat different fashion. However, the same germinal layers, though in a changed form, do make their

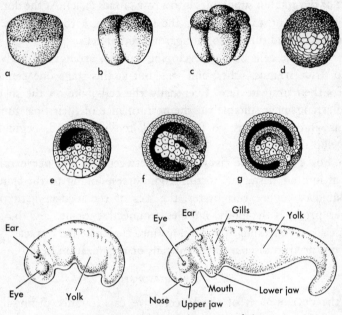

Fig. 73. Stages in the development of the newt.

appearance and produce the organs just as they do in the lancelet.

The egg of a bird contains such an enormous quantity of yolk that the yolkless protoplasm with its nucleus floats like a little droplet on its surface. (Compare Fig. 71; the egg is turned round to show the germinal vesicle in top view.) No wonder this small mass of protoplasm cannot manage to divide the entire quantity of yolk. The cleavage is restricted to the germinal vesicle from the rim of which the original gut develops inward (Fig. 74). Eventually all three germinal layers lie flat on top of the yolk. If we had not studied the simpler conditions in the lancelet, our chances of understanding the process would be slight. The rim of the germinal vesicle finally grows right around the yolk, which thus comes to lie inside the embryo, ready to be used up.

As in the lancelet, the mammalian egg has very little yolk, but for a different reason. The lancelet does not need much yolk

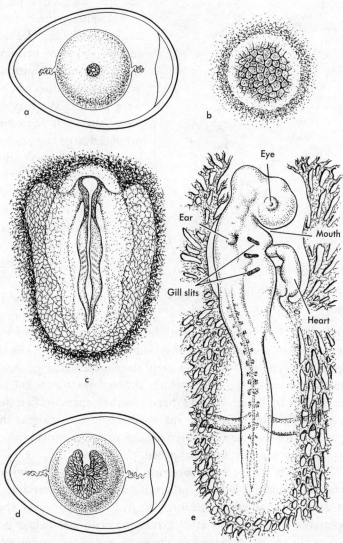

Fig. 74. Development of the chick.

because it starts feeding at a very early stage of its development;
the mammal, because it develops within the body of the mother

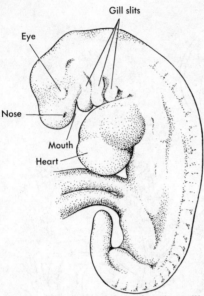

Fig. 75. Human embryo, showing gill slits.

and gets its food from it.

Despite the small amount of yolk, cleavage and formation of germinal layers in mammals proceed as if large quantities of yolk were present. We may assume that our ancestors produced large eggs, rich in yolk. In fact the most primitive of the now living mammals, the curious duck-billed platypus and the spiny ant-eaters of Australia, do not give birth to living young, but lay eggs with plenty of yolk.

This is one of the instances, and there are quite a few more, in which detailed study of the development of embryos gives us a clue to the early history of a species. Thus the human embryo (Fig. 75) at an early developmental stage shows something resembling gill slits, which, although nonfunctional, can be explained by the assumption that our earliest ancestors breathed through gills. This tallies with the opinion held by biologists that all vertebrates are descended from fishlike ancestors, however long ago this may possibly have happened.

In its adult stage every creature is adapted to the environment into which it is born, otherwise it would not survive. In the developmental stages, features of great antiquity make their appearance in generation after generation with astounding persistence. In the development of the large biological groups as in the development of the single individual, these recall the rungs of the ladder by which Nature reached its final stage. In the evolution of animals, as in other things in this life, it appears easier to use a beaten track over and over again than to make new pathways every time.

2. The Formative Forces in Development

When the hand of the artist shapes the form of a bird from a mere lump of clay we observe both cause and effect and everything is quite obvious. But how can an egg cell shape itself into a bird? Past generations of biologists rested content with observing and marveling at this process and with the discovery of the factors that the developmental processes of all animals have in common. But soon the question arose: Where is the hand of the modeling artist?

Unfolding or emergence?

There was a time when the easy answer seemed to be that the end product of creation was contained alive but dormant in the germ cell and needed only to unfold and grow in size. When, after the invention of the microscope, the sperm cells were discovered, they were taken to be the form-giving elements, the egg being only a formless mass serving as a nourishing medium during development.

Imagination carried investigators astray to such an extent that they believed they recognized in the swollen front part of a human sperm cell a tiny curled-up human being (Fig. 76). They failed to consider that an infinite number of smaller and still smaller human beings would have to be contained within the one they thought to see, representing generation on generation until the end of time. For if the future generations are not preformed in this way, the problem reappears again, though at a different point. The idea of a miniature creature unfolding by mere growth

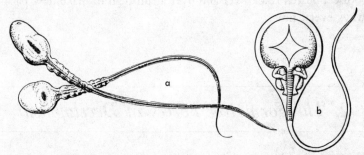

Fig. 76. Left, human sperm (enlarged 1000 times); right, an old idea of what might be in the sperm – a tiny curled-up human being.

was completely abandoned as soon as better microscopes revealed the structural details of germ cells. The early stages of development, which we discussed in the previous chapter, do in fact show that the shape and structure of each individual emerge anew in every case and are gradually built up.

Nowadays we put the question differently: Is the final fate of parts of the egg cell determined by its intracellular structure or will this be decided by the operation of formative forces during the process of development? Only experiments can give us the answer, and strangely enough it was found that both possibilities are realized.

Artificial twins

Although sea squirts resemble saclike or grapelike plants, they are considered to be rather lowly relatives of the animals from which vertebrates stem. There the development of the egg starts, as in all animals, by division into two cells. These look exactly alike and it is possible to separate them without damage. In fact they continue to develop, but each cell produces only half an animal. One cell produces the left, the other the right half of a sea squirt. Quite obviously the fate of the two halves was already determined at the first division. If we perform the same experiment with the egg of a newt the result is completely different.

Each of the two cells develops into a complete and intact animal,
but each animal is half the size of a
normal newt at the time of hatching, because the artificial twins have
to share the food resources of one
egg (Fig. 77). During normal development each of the two cells
would have formed one half of the
final animal, but after separation
each produced a whole animal. This
shows that the egg of the newt is
still quite freely adaptable at the
two-cell stage.

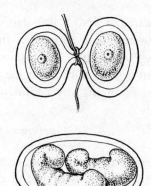
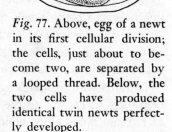

Fig. 77. Above, egg of a newt
in its first cellular division;
the cells, just about to become two, are separated by
a looped thread. Below, the
two cells have produced
identical twin newts perfectly developed.

In this experiment identical
twins are artificially produced.
Identical twins occur sometimes naturally in Man. This is very likely
also the result of halving of the eggs
at a very early stage of development, each half forming a normal
fully developed human being. The famous Siamese twins are less
fortunate, because they are incompletely separated. We can produce such malformations artificially if we do not separate the two
cells of a developing newt's egg completely, but tie the egg along
the first furrow of cleavage (Fig. 78). This can be done by using
a child's hair for the tying off.

But how is it that one and the same operation, namely the
halving of the embryo at the two-cell stage, can lead to two such
different results, producing two halves of an animal in the egg of
the sea squirt, and two normally shaped dwarfs from a newt's egg?

Well, if we halve a newt embryo at a considerably later stage
of its development we do not obtain twins but half-animals. At
some time the fate of the different parts of the embryo must
therefore become finally determined. The difference between the
egg of the sea squirt and that of the newt is not so great as it ap-

peared at first; it is merely a matter of the time lapse before the future fate of all parts of the embryo is irrevocably fixed. In the egg of the newt this happens relatively late. Is it possible to find out a little more about the when and how of this decision? It is by no means easy because the egg is only the size of a pinhead.

Operations under the microscope

Under the microscope we can observe how the egg develops into the spherical blastula and the cup-shaped gastrula, and how the mesoderm is formed. What we cannot see is which part of the single-layered blastula will turn into brain and which into skin. This development takes many days, and the migration and rearrangement of parts is so slow a process that we can observe it as little as the movement of the small hand of a clock.

The anatomist Walter Vogt thought of an attractive method to help the eye. He stained the cells of the living blastula with dots of red and blue colors, which penetrated into the cells without damaging them. Thus he was able to study at leisure the destination of the colored cells and found out which part would form the brain and which the skin. However, Vogt did not obtain any information concerning the timing of their assignment. This problem was solved by the ingenious experiments of the zoologist Hans Spemann.

He used two newt embryos in the blastula stage and cut out two small pieces in two different places. In one animal he took a piece from the zone that normally turns into the skin of the stomach region, in the other from the zone that gives rise to the brain. The future skin tissue of one animal was then put into the brain zone of the other and vice versa. This is easier said than done, and it takes great manual dexterity and especially fine instruments to perform such operations on objects so small and delicate. The necessary instruments could not be bought in any shop, and Spemann made them himself from drawn-out glass threads and tubes.

The experiment was successful, and within ten minutes the implanted pieces healed into place, a fact that can arouse the envy of any surgeon. The development continued and the healing was so perfect that the normal and the implanted tissues could not be distinguished. Of course the transplanted parts could have been marked by stain, but Spemann found a simpler and, for his purpose, even more perfect solution. Some newts have dark eggs and some have light eggs. If one takes, from a dark blastula, a piece of prospective brain and transplants it into the skin region

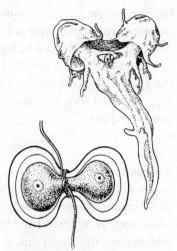

Fig. 78. Siamese twins of the newt produced experimentally by the incomplete tying off of the cells in the first division.

of a white one, the transplanted part will always be recognizable by its dark color (Fig. 79). The experiment showed that the piece of tissue that ought to have become brain developed into skin in its new place, according to its new position. Conversely the graft from the skin region turned into a bit of brain when implanted into brain region. Thus the grafted tissues adapted themselves in both cases to their new environment.

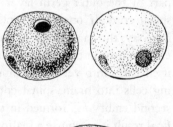

If the same experiment is made on slightly older eggs, for example at the gastrula stage, something different happens: the implanted graft develops accord-

Fig. 79. Experiments in transplantation in the newt.

ing to its origin. A piece from the brain develops, irrespective of its new environment, into brain right in the middle of the newt's belly skin. From this we conclude that the fate of the developing cells is sealed at the time when the one-layered blastula turns into the double-layered gastrula. The question of the "when" is therefore answered. But it is more difficult to find an answer to the question: How?

The "organizer"

Spemann made a strange and wonderful discovery. He and his pupils did the grafting experiments that we have just described in different parts of the body. It was found that one special part of the embryo differed from the rest. It is that part of the blastula that lies on the upper lip of the opening into the developing gastrula. This, you will recall, is known as the blastopore. The cells that lie there are, at an early stage, determined to become vertebral column and musculature, and no transplantation into any other region can change this.

But not only do they insist on going their own way, they also dominate the fate of neighboring cells. When the blastula develops normally, the lip of the blastopore is pushed under that part of the outer germ layer that will later form the brain and spinal cord. If one transplants material from the blastopore lip into, say, the flank of the blastula, it influences the cells of its new environment in the way it would in its normal position. The graft itself turns into vertebral column and musculature, the neighboring cells into brain, spinal cord, eyes, ears, and nose; in short, a second embryo is formed in the side of the first (Fig. 80). The final result is of course a malformation. But it shows that there are certain regions in the embryo that have form-giving powers, and Spemann called them organizers. The positions of such organizers can be retraced to the egg cell.

What is it that turns a restricted zone of the embryo into an organizer? Is this a matter of structure in the living protoplasm, or are there substances of a chemical nature that exert their influence on their surroundings?

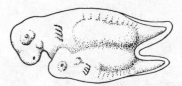

Fig. 80. An embryo of the newt, onto which the "organizer" (material from the blastopore lip) has been grafted, thus causing another embryo to develop below.

The latter seems to be case. This is borne out by the surprising result of an experiment tried by Spemann and his pupils. If one cuts a piece of skin from a blastula and grafts it near the organizer of another embryo, it in turn becomes an organizer and, grafted into another blastula, will produce an additional embryo. This points to chemical substances having been taken up by the transplanted tissue from the organizer and handed on to its new surroundings.

Other experiments too point in this direction. Hardly any other tissue has been tormented so much as the upper lip of the blastopore of the newt embryo. It has been boiled and frozen; it has been pickled in alcohol and ether, in acids and all kinds of fluids; it has been crushed most carefully and the structure of its cells destroyed. After all such procedures it was found that the much maltreated material showed itself extremely resistant. It continues to function even in the form of a shapeless mash. All this therefore speaks in favor of the chemical nature of the organizer.

Still further clues were obtained when a piece from the prospective brain region of the frog embryo was grafted into the prospective mouth region of a newt. It formed a mouth in accordance with its new environment; it was not, however, the toothed mouth of a newt but the mouth of a frog, with its characteristic horny jaws. This means that the organizer determines what the parts are going to develop into to form a co-ordinated, functional unit. What the parts finally look like depends on their origin.

Spemann was honored with the highest international distinction for scientific achievements. He received the Nobel Prize for his discovery of the organizer and for having lifted the first veil of the mystery that surrounds the formative forces at work in the development of animals.

3. Growing Up by Transformation

The significance of a larval life

As children we observed with wonder how the wormlike caterpillar changes into a pupa, and how from the pupa emerges the winged, resplendent butterfly — a pleasure not unlike the one we felt when watching a conjuror's tricks. The thoughtful biologist is interested in the significance of this transformation. Why does the chicken egg produce a little chick differing from the hen chiefly in size, while from the egg of a cabbage white emerges a caterpillar different in every respect from the adult butterfly?

We know that insects, crustaceans, and all arthropods have an armor of chitin that hinders the growth of their body. They therefore molt from time to time by throwing off their armor. The wings too are covered with chitin. But their size and shape would make it impossible for the insect to withdraw from them during molting. As long as insects grow they have therefore no wings or only short thick stumps, out of which they can molt without trouble. Flightlessness makes them adopt a different mode of life, and there are a number of important differences between the winged adult and the young insect.

A progressive businessman who wants to make alterations to his premises has to make up his mind whether he can keep his shop open during alterations or whether he has to close down for a time. This depends on the amount of alteration involved. Insects behave similarly. Young grasshoppers live very much like the adults and look very much like them, except for their undeveloped wings. Here the changes can be made without interrupting the usual business. With the last molt the wings grow to their

Fig. 81. Metamorphosis of the grasshopper.

full size and the final adult stage is reached (Fig. 81). The same happens in aphids, bugs, and many other insects. A caterpillar, on the other hand, is so different from a butterfly that during the alterations business may seem to close down. A pupal stage intervenes between caterpillar and butterfly. Appearing from outside to be in a stage of rest, the pupa is a hive of activity inside. The organs of the caterpillar are being demolished and its shape altered into that of the butterfly, which in the last molt emerges from the chitinous cover of the pupa. A similar sequence of events occurs in beetles (Fig. 82), flies, bees, and many other insects.

The youthful stages, which differ essentially in shape from the adult ones, are generally called larvae. The caterpillar is the larva of the butterfly, the tadpole the larva of the frog. A frog has neither wings nor a chitinous skeleton. In this case there must be another reason for a change of shape.

All life began in water. To live on land demands special adaptations. Frogs, like the rest of the amphibia, are not quite perfectly adapted for a life on land. They have a very thin skin and therefore are in need of moisture. Moreover, their delicate, small young would dry up on land in no time at all. They live in water as did their fishlike ancestors, and they have fins and gills. Only when they have grown larger and tougher and more resistant do they lose their gills and tail. They grow legs and turn into proper frogs, which can live on dry land as well as in water.

The butterfly grows wings, the frog grows legs; however, transformation does not always represent an advance in organization. Earlier, on page 156, we described crustaceans which as adults live a parasitic life on fishes. Their larvae hatch from their eggs as perfectly normal, charming little crustaceans, equipped with legs and eyes and swimming about freely in the water. Only after

Fig. 82. Metamorphosis of the Japanese beetle.

they have attached themselves to a fish do they turn into shape-less sacs, without legs and sense organs, drawing food from their host, only to produce a large quantity of eggs.

The transformations in the three examples discussed seem to differ widely in their significance. But they have one thing in common. The winged insect takes to the air, the amphibian con-quers the dry land, and the parasitic crustacean acquires a com-fortable existence. As adults they are specialists, as youngsters and larvae they live in the way typical for their class.

Larval stages are quite common in animals, and they exist for a great variety of reasons. Thus the eggs of sponges, jellyfish, worms, snails, starfish, and sea urchins give rise to larvae.

Sea urchins produce an enormous number of small eggs with little yolk, and their young have therefore to find food at a very early stage of development before they have had time to reach their final shape. The marine sponges live a sessile life on the sea bed. They would soon be very much on top of each other if all their eggs developed right where they were laid. As it is, ciliated larvae hatch from them, which can swim away to find new living space. One speaks of larval dispersal in such a case. The lobster, held in high esteem by gourmets, is found at the bottom of the sea, and its larvae swim in the open ocean. Nobody would guess that the larvae, which remain suspended in mid-water thanks to their broad, leaflike shape, are nothing but young lobsters.

4. *The Tending of the Young*

Maternal care

The simplest kind of maternal care involves nothing but the laying of eggs in a suitable place for the hatching young brood. The cabbage white, which lays its eggs on the right food plant, provides for the further development of the young caterpillars. Hunger, however, is not the only source of danger to a young life. Eggs and newly hatched animals are coveted morsels, which are eagerly preyed upon if they are unduly exposed. In the majority of cases the mother no longer cares for the young after the eggs are laid. Sometimes, however, especially in insects, the techniques employed for the provision of food and sufficient protection are astonishing and call for complicated instinctive actions.

We have heard how the solitary leaf-cutting bee builds for each egg a nest made of specially cut pieces of leaves. Even more extraordinary is the mason bee which seeks an empty snail's house for each of its eggs. Into this it first deposits a store of pollen and nectar, laying its egg on top of this honey bread, so that there is food in plenty. But this is not all: the mother bee protects her nursery with all the means at her disposal against any robbers likely to eat the honey or the fat, helpless larva. First she seals off the brood chamber by a wall, made from chewed leaves and hard-setting saliva. The whole of the remaining space of the snail's house is then filled with little stones, collected singly and laboriously. To prevent this barricade from rolling out, a second wall of leaf cement is erected. Finally a protective grass roof is constructed above the snail's house which completely hides it from view. Wrapped up like an Egyptian mummy, the larva develops and pupates. When it has turned into a winged bee, it struggles from its isolation into the sunlight.

Devoted parents and happy-go-lucky parenthood

The adult mason bee lives for only a few weeks and has a small number of offspring. It has no time to make a great number of elaborate nests. It is easier for a cabbage white to be fertile, as its parental commitments are limited to laying the eggs on cabbage leaves. On the other hand, the numerous caterpillars are extensively hunted and perish in much greater numbers than bee larvae. As a rule the less care is taken for the young, the greater the number of eggs, and in this way an average level of population and the survival of the species are guaranteed. This is convincingly demonstrated in animal groups in which we find, side by side, devoted and very casual mothers.

No parental care is give to the majority of fish. The parents have done their share once they have discharged their germ cells into the water. Again, what is lost by lack of care is made up by sheer numbers. A single sturgeon produces in one go three to six million eggs, the caviar of the gourmet. The carp lays up to half a million eggs, but our stickleback hardly a hundred. Among its kind it is the exception, and very significantly it builds a nest.

So far we have talked only of devoted mothers. In the stickleback it is the father who builds the nest and looks after the young. The nest is elaborately constructed from bits of stems, leaves, and roots, which by means of a sticky secretion are glued into a shallow cavity in the sand. The nest is defended and approaching rival males are viciously attacked. When all is ready the male goes in search of a female, and if the chosen one does not follow willingly he can be seen to bite and push her in a seemingly unchivalrous manner. When the female has laid the eggs, he fertilizes them at once and from then on defends them with fury and perseverance against any intruder who dares to come near, be it the mother herself. Even after hatching, the young fish are looked after by the father and every time they try to swim away he catches them with his mouth and spits them back into the nest. This goes on for several days, until they finally disperse and go their own way.

The snapping up and spitting back of the young is quite a common habit among nest-building fish. From this behavior may have developed the habit of some species of carrying even the newly laid eggs about in their mouth until they hatch, and even the young after hatching. That the parents refrain from swallowing their young is all the more astonishing since the weeks taken up by the care for the brood mean a period of hunger for the old, which, having a mouth full of children, cannot eat a bite themselves. It has been observed that after having left the mouth of the parents, the young fish at first stay close to them, taking refuge in the paternal mouth at the slightest sign of danger.

Among frogs and toads we come across the same correlation between the number of eggs and parental care. Most toads produce about 10,000 eggs and leave them to their fate. But Pipa vulgaris, a toad, produces only about forty to sixty eggs, and here the male helps the female to put each egg separately into folds in the skin of her back, which looks like a honeycomb. The eggs develop into tadpoles in their separate cells, staying there until they have changed into little toads (Fig. 83).

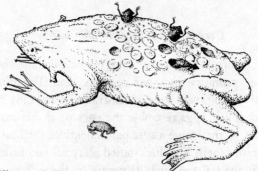

Fig. 83. The toad, Pipa vulgaris, with young in the "pockets" of the skin on its back.

There are cases of postnatal or true care of the young in which the young are not only provided for during their development but cared for after leaving the egg. In lower animals and cold-blooded vertebrates such cases are the exception. Among the warm-blooded birds and mammals extensive care of the young is the rule. The fundamental difference between them lies in the fact that the eggs of birds develop outside, those of mammals inside, the maternal body.

Birds' eggs are large because they contain a lot of food. The female incubates them in a nest, and the hatching young are for a longer or shorter period cared for by the parents. In mammals the care of the young reaches its highest level, because their small eggs, containing little yolk, develop within the body of the mother and are nourished by it. When born, the young feed for weeks or months on the milk of the mother, which is easily digestible and always available. Yet neither the production of living young nor the feeding of them on milk is an "invention" of the mammals. Some worms, for example, produce living young. Some flies, living on meat, lay fully developed larvae. In certain fish, snakes, and in a few other animals, we find that the embryos develop inside the mother. Furthermore, we have already seen that the larvae of bees are fed during the first days of their life on a glandular secretion, which is in kind and origin comparable to the milk of mammals.

Care of the young in mammals

Only a few mammals are an exception to the rule, and they strike one as rather primitive in other respects also. The duck-billed platypus and the spiny anteater of Australia are the only mammals that lay eggs. Their eggs resemble the eggs of snakes rather than those of birds, as they have a soft parchmentlike shell rather than a hard calcareous shell. The duck-billed platypus incubates its eggs in a hollow in the soil and feeds its young on the secretion of milk glands. The spiny anteater lays only one egg, which it pushes into a skin pouch on its belly, where the young spends the first days of its life in cozy comfort feeding on milk.

The marsupials, of which the kangaroo is the best known, have brood pouches. The egg develops within the female body, but the placenta, through which in all other mammals the developing embryo obtains nourishment, is not perfected. Therefore the embryos are born in a premature and helpless stage. The kangaroo, itself larger than a man, gives birth to a young as small as a walnut. The newborn crawl into the pouch of the mother, where they become attached to the nipples of the milk glands and continue their further development. Even later, when the young can jump about independently, it hurries back into the pouch of its mother in case of danger.

In all other mammals the embryos complete their development inside the womb. As a rule, once a year a few eggs are released from the ovary and reach the oviduct. In Man this process is not seasonal but occurs regularly every four weeks. A strange change occurs in that part of the ovary from which the egg was liberated. The vacated space fills up with tissue, the so-called yellow body. This produces a hormone that influences the further course of pregnancy. The hormone produced travels in the blood stream all through the body and eventually reaches the womb or uterus, the mucous membranes of which get stimulated into active growth. Its blood supply increases and other changes occur, which prepare it to receive the approaching egg. The journey through the oviduct takes several days. The hormone has announced its arrival and has initiated the necessary arrangements for its reception.

The oviduct is the only place in which the egg can be fertilized. If fertilization does not occur, the egg perishes. The yellow body disappears, the womb remains unoccupied, and its specially prepared inner membrane is shed. The building up and consequent shedding of the inner lining of the uterus happens periodically, once a month. It is known as the period or menstruation (from the Latin *mensis* — month). If, however, the egg has been fertilized, it attaches itself closely to the wall of the womb and starts its development there.

The growing embryo forms an outer cover that implants itself with rootlike appendages in the maternal membrane (Fig. 84). Thus it enters into a most intimate connection with the maternal tissues, which together with rootlike appendages from the embryo form the placenta. In Man this has the shape of a disk, in other mammals it can have different shapes. There is, however, no true exchange of blood between mother and embryo. The blood vessels of the maternal membrane and those of the developing embryo are separated from one another by very thin sheets of tissue, which allow a lively exchange of various substances. Thus the placenta is responsible for the nutrition and respiration of the fetus until it is born, when by the severance of the umbilical cord the newborn and the placenta part company for good. The timely secretion of milk is also brought about by the action of a hormone.

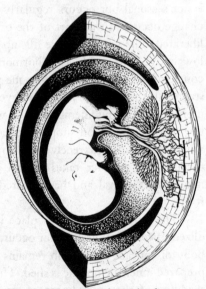

The number of young in a litter and the duration of pregnancy vary in different mammals. A mouse produces its litter in three weeks, a rabbit in four. In cattle and in Man the duration of pregnancy is nine months, in the horse eleven, and in the elephant twenty-two. As a rule it increases with the bulk of the body, but there is no strict proportionality. All newborn mammals, however, need to be fed and cared for by the mother.

Fig. 84. Human embryo in the uterus; right, the placenta. The blood vessels are in black.

In Man this care extends far beyond the requirements of the body. This kind of care is unknown in any, even in the highest, fellow mammals. Nevertheless elephants, bears, and apes occasionally shake their youngsters or box their ears if they do not "behave" themselves, and it looks as if the educational efforts of human parents had their roots as early as in the animal kingdom.

VI

HEREDITY

1. Something Obvious, Yet Worth Pondering

What a stir it would cause if a stork emerged from a chicken egg! Biologists all over the world would put their heads together. But nothing of the kind has ever happened. The egg of a chicken produces a chick, and the stork egg a stork. However obvious this may appear, let us consider what it means.

Father and mother both contribute

The stork egg is a cell from its mother's body. That this cell, once fertilized, develops into a bird that grows black and white feathers in a special pattern, has tall legs equipped for wading and a long red beak, feeds on frogs, is gripped by wanderlust when August comes and makes off for Africa — in short, the fact that this cell produces yet another stork, with all its characteristics of body and behavior, must all be the doing of the fertilized egg cell. All happens in the same way if one incubates the egg artificially and the young stork never sees its mother. There remain the two germ cells, the egg cell and the sperm cell, as the only links that can be made responsible for the similarities.

As we know, the formative powers are contained in a very small portion of the egg's bulk, namely in a nucleus, surrounded by the little drop of protoplasm. Those who find this impossible

220

to believe need only consider that the human egg or the egg of a lancelet, although it is almost microscopically small, always reproduces its kind.

That the progeny should as a rule resemble each parent to some degree may still be surprising, because the egg is so much larger than the sperm. Even in Man, where the egg is relatively small, measuring little more than 1/100th of an inch across, 200,000 sperm cells occupy the space of one single egg. The difference in size, however, concerns only the cell protoplasm. Under similar conditions the nucleus of a sperm is as large as the nucleus of the egg and contains the same number of chromosomes. It therefore seems obvious that the nucleus is responsible for the characteristics of the parents, a conclusion confirmed in every detail by science.

At marriage we can predict with certainty that the children of a young couple will be of human shape. However, whether the nose of the offspring will be like its mother's or its father's, or like the nose of one or another of its more remote ancestors, remains to be seen; there seem to be no set rules for predictions of this kind in heredity. But it certainly is worth while to find out whether heredity is after all governed by rule or law.

2. A Monastery Garden in Brünn

The convenience of experiments on plants and animals

Good ideas are often very simple, and it was such a good idea that made Mendel cross two plants that differed in just one easily recognizable hereditary character.

The use of carefully chosen animal or plant material has the advantage that one can plan and control breeding experiments and within a short time produce many generations of numerous offspring. It takes fourteen years, at least, for a woman to become mature, but the little fruit fly, the most thoroughly studied animal in the science of heredity, can produce a new generation of offspring every fortnight.

You may say that for Man heredity in flies and plants can be of only secondary importance. However, it was found that in heredity, as in all biology, certain fundamental processes are common to all living cells, be they plant, animal, or human. Thus up to this day the experiments on animals and plants laid the foundation for all our knowledge concerning human heredity.

The forgotten publication

The first experiments on the crossing or hybridization of plant varieties, carried out with the explicit intent to study possible laws of heredity, were undertaken in the middle of the last century by a monk. In the garden of a monastery at Brünn, Gregor Mendel grew purplish- and white-flowering pea plants. He carefully kept insects from them and himself transferred the pollen from white to purplish flowers, studying the colors of the flowers in the next and subsequent generations. The results of these and similar experiments, which were carried out for a number of years in the quiet kitchen garden of the monastery, were published in 1865 in a treatise, which at that time nobody understood and to which therefore nobody paid any attention.

A quarter of a century later similar experiments were carried out by others. By mere chance, they found to their astonishment that what they had thought to be a new discovery was already set forth in Mendel's publication. Mendel had not only been a most careful observer but he had interpreted his results with astonishing ingenuity. By that time he was dead, but as he was the first to have recognized and formulated the strictly recurring and repeatable events in hybridization, these findings were called Mendelian laws, in his honor. They are the foundation on which modern genetics developed rapidly into a branch of biology in its own right.

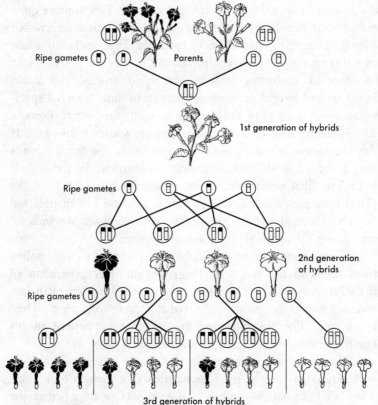

Ripe gametes   Parents

1st generation of hybrids

Ripe gametes

2nd generation of hybrids

Ripe gametes

3rd generation of hybrids

Fig. 85. Cross of red and white four-o'clocks.

Now you will want to know what these laws are. We shall discuss them presently, though we shall not present our examples in their historical sequence but choose experiments, old and new, that are most suitable to further our understanding.

Flowers and guinea pigs obey the same laws

The common four-o'clock occurs as a red- and as a white-flowering variety. If these are crossed and the resulting seeds sown out, one gets a generation with exclusively pink flowers. They are hybrids, with a color intermediate between the white and the red of the parents (Fig. 85, first hybrid generation). This appears quite plausible. If members of this first hybrid generation are crossed among themselves, with insects carefully kept away to make sure that no outside pollen is introduced, we do not get, as might be expected, uniformly pink flowers again. Instead, we obtain in the second hybrid generation, apart from pink-flowered speci-mens, some with pure red and some with pure white flowers. The characteristics of the grandparents are showing up again. If this experiment is carried out with many seeds, the distribution is seen to occur in definite numerical proportions. In the second hybrid or filial generation, as the geneticists put it, half of the plants have pink flowers, one quarter white, and one quarter red flowers. (Second hybrid generation: to save space we indicate each group by one single representative flower.)

If one continues to pollinate each plant only with pollen from plants of its own type, in generation after generation all the white-flowering plants yield only white-flowering offspring, and red-flowering plants only red-flowering offspring. They breed true. The pink-flowered group, however, behave always like the first hybrid generation just described (Fig. 85).

The same happens in animals: red-haired guinea pigs crossed with white-haired ones produce hybrids all of which look alike. They all have a pale reddish coat, irrespective of whether the father was red and the mother white or vice versa. The second hybrid generation again splits up in the same numerical propor-tion as the flowers of the four-o'clock.

If two pale reddish individuals are crossed only once, we cannot, in the relatively small litter, expect to get consistent numer-ical proportions. If the experiment is repeated a hundred or a thousand times the results become statistically reliable, and we get a distribution of a definite ratio of different offspring. We are familiar with chance events. If one puts one thousand black and one thousand white marbles into a sack and shakes them well, one has an equal chance of picking a white or a black marble at any given instant. If one takes two at a time both might be black, or both white, or we might pick a black and a white one. The result will be a matter of mere chance. However, if we take out a hundred it is reasonable to expect that we shall pick approximately fifty black and fifty white ones. The more often we repeat this the nearer we get to average numerical proportions. If the laws of chance did not operate, betting could not exist, nor any insurance companies. Both depend on sufficiently large numbers to be able to rely on the laws of chance, as if they were a law of Nature.

For the moment we conclude that in our breeding experiments chance must have a finger in the pie, as the ratios become more reliably accurate the more often we repeat the experiment.

Mendel's experiments on peas

The cross between purplish- and white-flowering pea plants that Mendel carried out gave a slightly different result from those discussed so far. The first hybrid or filial generation consisted of purple-flowering pea plants only (Fig. 86). In this case the purple color asserted itself and geneticists say it was shown to be the dominant factor. The factor for white color got suppressed and is therefore called recessive. If, however, one crosses the first hybrid generation with others of the same generation (selfs them) it becomes obvious that the factor for white had remained present all the time, though hidden. It happens that the second hybrid generation produces about three quarters of plants with purple flowers, while one quarter of the progeny has white flowers. The latter all breed true but only some of the purple-flowering plants are true breeders, whereas two thirds of them split up again. In animals too we can find examples that behave in the same way. One important fact has emerged. The hereditary make-up can contain factors that need not manifest themselves externally.

Despite certain differences between the experiment on peas and on the four-o'clock, two rules are quite strikingly obvious: the members of the first filial generation are all alike in each case,

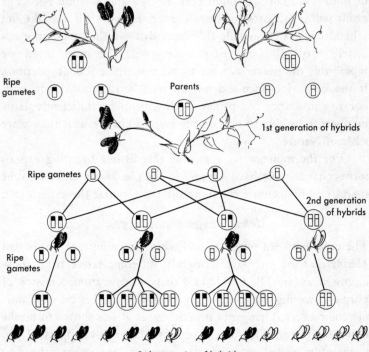

Fig. 86. Cross of purple and white garden peas.

and the second filial generation splits up according to a definite law of chance.

Mendel had interpreted his findings correctly. He believed that the cause for the appearance of a certain color in a flower is what he called a definite hereditary factor. This factor must be contained in the germ cells, as they alone form the material link from generation to generation. In order to explain the extraordinary splitting up of these factors in the second filial generation, he made certain assumptions about the behavior of hereditary factors in the germ cells and in the body cells that illuminate his extraordinary power of reasoning and that were most brilliantly confirmed by later scientific investigations. We shall soon hear what actually happens.

Considering what different inherited qualities, intelligence, and incurable disease mean for people, we realize that the insight into the laws of heredity was a discovery of the greatest importance. There is a monument in honor of Mendel at Brünn.

3. *Two Lines of Research Meet*

There are many different approaches to biology. Some biologists like to explore the secrets of primeval forests; others are attracted by the risk and excitement of experimentation. Some like to study elephants, and others prefer to look at minute organisms, visible only under a microscope. Since the invention of the microscope, the study of the internal structure of germ cells, one of Nature's greatest riddles, has become one of the most fascinating of challenges.

On the face of it a young germ cell looks exactly like any other cell. All we see is a little lump of protoplasm and a nucleus. The latter is obviously of special importance in the process of inheritance. We have already mentioned that the egg and sperm cells, which fuse with one another during fertilization, have nuclei of equal size but differ in their quantity of protoplasm. If paternal and maternal characteristics are inherited in equal measure this must be due to those equal-sized nuclei. We can forget about the clear nuclear sap, but let us look closely at the chromosomes, which are carefully duplicated when they are passed on to other body cells. This surely points to their very special importance.

In the resting nucleus we cannot find out much about the

Fig. 87. Left, chromosomes of the fruit fly, Drosophila. Right, chromosomes of Man.

chromosomes, for their arrangement is not discernible. It is during cell division that we can observe several remarkable things, provided we have a good microscope. The whole nucleus and more so its very small inclusions in its interior are to the unaided eye far beyond the limit of visibility. If we magnified a poppy seed in the same proportion as we have magnified the chromosomes drawn in Fig. 87 it would have a diameter of four feet.

Chromosomes as personalities

First of all we find that the number of chromosomes differs in different species of animals and plants. In the fruit fly each dividing nucleus contains eight chromosomes, in the mouse forty, in Man forty-six, in certain butterflies sixty-two, and in a roundworm only four. The number of chromosomes in each species, however, is always the same, whether one studies a dividing skin or gland cell, or a germ cell. However small, chromosomes seem

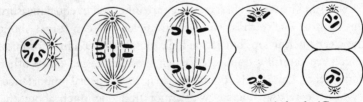

Fig. 88. Maturation division. The chromosomes are halved. (Compare with Fig. 64.)

to be personalities each with its characteristic shape and size. Not only their number but their shape too remains the same. The following fact is most remarkable: if one draws a careful copy of all the chromosomes of a nucleus and arranges them according to shape, one always gets pairs of equal size and form (Fig. 87). Each individual chromosome is therefore represented twice; the nucleus thus contains sets of chromosome pairs.

It is not difficult to guess how this comes about. During fertilization the nucleus of the egg fuses with the nucleus of the sperm cell. Each has its own set of chromosomes. Observation teaches us that chromosomes keep their identity and exist side by side. One set of chromosomes therefore comes from the nucleus of the egg, the other set from the nucleus of the sperm cell. Since in all subsequent cell and nuclear divisions, during the development of the body, the chromosomes always duplicate themselves and are equally distributed between the daughter cells, every cell, to whatever organ it belongs, receives the same two complete sets of chromosomes. Thus the two sets of chromosomes reach the young germ cells too. These are often set aside at a very early stage and are destined to give rise to a new generation later.

Now it looks on the surface of things as if the number of chromosomes were bound to become doubled every time an egg fuses with a sperm. But that cannot be. If it had started with our earliest ancestors, we should by now have accumulated many more than forty-six chromosomes!

A careful observation of the germ cells offers us the solution to this problem. Before fertilization a special type of cell division occurs both in the egg and in the sperm cell, during which the number of chromosomes is reduced by half. This division is called the maturation division (Fig. 88). Fertilization then restores the characteristic chromosome number.

The maturation division of the germ cells

The maturation of germ cells is rather a complicated matter. But if we leave aside all the unessential details, we can explain the essential features briefly.

To simplify matters we choose an animal with a small number, say six, that is twice three, chromosomes in its body cells (Fig. 88). The cell destined to give rise to the spermatozoa has also twice three chromosomes in its nucleus, as do all other body cells. In an as yet unexplained way the chromosomes come to lie together in pairs. Every chromosome from the paternal set pairs off with the equally shaped and sized chromosome of the maternal set. A spindle is now formed just as in ordinary cell division. However, this time whole chromosomes are pulled apart from one another, and as they were arranged in pairs of equals, each daughter nucleus receives a complete set of three chromosomes. The same happens in the egg. In fertilization the nucleus of the egg and that of the sperm cell unite to form the fusion nucleus with six chromosomes, the number characteristic for the species.

So far the study of cells and that of heredity had gone their separate ways. Scientists soon realized that these two fields linked up. They pointed out that the findings of Mendel, as expressed in his laws of heredity, can be explained without difficulty by the events observed during maturation division and fertilization, if it is assumed that the hereditary factors that determine the make-up of an organism lie in the chromosomes. To elaborate this let us use one of the breeding experiments discussed earlier on p. 224. It does not matter whether we choose our example from the animal or plant kingdom, because the behavior of the chromosomes at maturation division and fertilization is essentially the same in plants and animals and Man.

The behavior of the chromosomes explains the results of hybridization experiments

We assume that the red color in the flower of the four-o'clock is due to a hereditary factor, which is present twice and contained in two separate but similar chromosomes. We indicate it by two black square dots in the diagram of our chromosomes in Fig. 85. The other chromosomes are left out in our drawing because they do not count for our purposes. The hereditary factor for red color is present twice, once in each one of the two corresponding chromosomes, because they are derived from two red-flowering parents. In the chromosomes of the white-flowering variety the hereditary factor for white color lies in the same place on the corresponding chromosomes (Fig. 85, under drawing to the right).

After the maturation division each germ cell contains only one set of chromosomes and therefore only one of the two chromosomes containing the color factor. The diagram (Fig. 85) shows a body cell near each parent and underneath the two ripe germ cells. During fertilization the nuclei that contain the red and the white factors fuse. Therefore the hybrid now has in all its body cells, which originate from the fertilized egg cell, two chromosomes with different color factors. Both together are responsible for the final color of the flower, and this is pink. Now it is obvious why all individuals of the first filial generation are alike in color: they all contain the same hereditary factors.

When the germ cells of the hybrids, the first filial generation, ripen, corresponding chromosomes pair off as usual and are separated during maturation division. Thus we get male and female germ cells, equal numbers of which contain the red factor or the white factor.

During fertilization four combinations are possible: a sperm cell with the red factor can meet an egg cell with the red factor; two, a sperm cell with the red factor can meet an egg with the white factor; three, a sperm cell with the white factor can meet an egg with the red factor; four, a sperm cell with the white factor can meet an egg with the white factor. The first combination results in red-flowering individuals that breed true red, because no factor for white color is contained in their fused nucleus. The second and third combinations produce hybrids that will again split up, while the fourth combination will produce true-breeding white plants. It is a matter of mere chance which combination happens to come about. Since the two kinds of germ cells are present in equal numbers, there is an equal chance for any of the four combinations to come about. Now we understand the role that chance plays, and that with a sufficiently large number of offspring we are bound to get, according to the laws of probability, true-breeding red plants, pink hybrids, and true-breeding white plants, in a ratio of $1:2:1$.

The different results in pea plants can be quite easily explained in the same way. The distribution of the hereditary factors is exactly the same, only their effect in the offspring is different. In peas the purple color of the flower is dominant. Now we understand what this means. The first hybrid generation contains the factors for color of both parents. However, the factor for white does not assert itself. The hereditary factor for purple color is the dominant one. All hybrid flowers are purple. For the same reason, in the second filial generation we shall get, besides true-breeding purple plants, hybrids with purple flowers; only in true-breeding white plants, which contain no factor for purple, does the recessive factor for white color show up. Therefore three quarters of the offspring have purple flowers. The purple flowers are entirely alike in their external appearance, but they differ in their hereditary make-up. This shows up only on further crossing. The true breeders that have in both corresponding chromosomes the factor for purple will go on breeding true, while the hybrid purples will, for the same reasons as the first hybrid generation, give rise to a certain proportion of white-flowering plants.

The multiplicity of hereditary factors and their arrangement on the chromosomes

We have studied one single hereditary factor only. The same chromosome on which we assumed this factor to lie contains many other hereditary factors, and they seem to be arranged in a row, one behind the other. All the other chromosomes contain

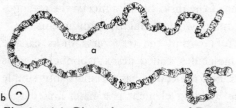

Fig. 89. (a) Giant chromosomes of Drosophila; the dark disks represent the hereditary factors; (b) an ordinary chromosome of this fly shown at the same magnification.

hereditary factors in a similar arrangement. In the fruit fly (Drosophila), one of the most thoroughly investigated animals, the number of hereditary factors is estimated to be about 10,000. We know exactly what effect many of them have. Morover, we know precisely in which chromosome they lie and what place they occupy on it. If one considers that the chromosome of a fruit fly is only 1/8,000th of an inch long, and that along this minute stretch several hundred hereditary factors are aligned, one cannot expect even the best microscope to show them up.

In the especially large nuclei of the salivary gland of Drosophila, and in those of midge larvae, are found giant chromosomes 100 to 200 times the size of the ordinary ones in other body cells. No one as yet knows why Nature should in this case exhibit the chromosomes in such a spectacular fashion. If, as a rule, she were not rather inclined to keep things under her hat, one might almost believe that this was a special arrangement to show the curious biologists the contents of these tiny treasure chests of heredity in a hundredfold enlargement. With the help of the most powerful lenses of a good microscope we see that the giant chromosomes are made up of disks of various thickness (Fig. 89). We know that each visible transverse disk harbors in some way specific hereditary factors. That this is correct has been convincingly shown by ingenious breeding experiments counterchecked by microscopic investigations. Malformations in certain races of flies, which point toward the lack of certain hereditary factors, were indeed connected with the absence of certain transverse disks in exactly that spot on the chromosome where the geneticist had foretold or expected its presence. The microscopically

recognizable arrangement of the disks fits to a nicety the arrangement of hereditary factors as worked out by experiments.

The hereditary factors within the transverse disks can be
traced to specific large molecules, called desoxyribonucleic acids
(DNA). Each of these DNA molecules has a remarkably simple
structure (Fig. 90). The building element is a base, linked to a
sugar, which in turn is linked to a phosphorus atom. The base
may be one of four possible alternatives: adenine (A), guanine
(G), cytosine (C), or thymine (T). The sugar is always desoxyribose, which together with the phosphorus provides the link to
the next building element. A coiled row of many thousands of
such building elements is one strand of the DNA molecule. Facing
this a second complementary strand is found in each DNA molecule. Through its bases it is connected with the bases of the first
strand in a very specific fashion, in which only the base-pairing
A-T and G-C can occur. Within a strand the sequence of the bases
determines the hereditary code, much like the simple Morse code:
in the latter the combination dash-dot-space carries the information
for the alphabet; in the DNA molecule combinations of the different types of successive bases carry genetic information. By means
of such a "message" a cell is informed how to construct the many
hereditary traits. Before the cell divides, the DNA molecule du-
plicates itself. The two DNA strands separate and while they do
so two new double strands are built up by some sort of zippering
mechanism. Hence all our hereditary traits such as the color of
our eyes, the type of blood we have, the basic mental elements
of our personality, the shape of our body, and a multitude of other
fundamental things are determined by specific code letters of our
chromosomes. And these code letters are nothing but specific base
sequences in the DNA. So Nature's greatest secret, heredity, is
hidden in just one type of molecule, constructed in a most simple
fashion.

The fertilized egg cell, with its double set of chromosomes,
contains most hereditary factors twice. This is generally true of
every hereditary factor but is not found in the X chromosomes
(p. 246). But for our present discussion we may assume that the

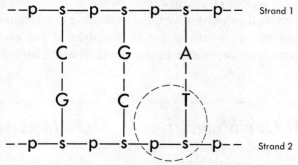

Fig. 90. Schematic drawing of part of a DNA double strand. In each strand the bases adenine (A), guanine (G), cytosine (C), and thymine (T) are connected by means of sugar (S) desoxyribose and a phosphorus atom (P). The two strands are complementary in that only the base pairs A-T and G-C can occur. Encircled is a single building element.

hereditary factors are doubled in one fertilized egg cell. Just as in our example the factor for color, so every other hereditary factor is arranged in a fixed sequence on a paternal as well as on a maternal chromosome. That means that the single set of chromosomes, as it is found in ripe germ cells before fertilization, already contains all the hereditary factors that are essential for the development of a normal individual. That this is so can be demonstrated by experiments.

After the completion of the maturation division one can stimulate the unfertilized egg of a sea urchin by artificial means to start its development. On the other hand one can let a sperm cell enter a fragment of an egg that is without nucleus. In both cases normal animals develop, although only half the set of chromosomes is present in each, in one case the maternal, in the other the paternal set. If, however, in that one set of chromosomes a single chromosome were missing we should get malformations, because of the complete absence of a group of hereditary factors.

Without the painstaking work on cells by the cytologists who tried to explore the structure of the nucleus in its finest details, the science of genetics could never have achieved what it did. Without the specialist we make no decisive progress in science and if he in turn is careful not to lose sight of the whole, his efforts will never become sterile through narrow and limited vision.

4. *Our Studies Become More Involved*

So far we have discussed only hereditary factors that affect the color of organisms. There are many others, such as those that control the growth of the body, the length of ears, the curling of hair, the number of fingers, the proneness to certain illnesses, musical talent, and innumerable other hereditary traits. If we tried to conjecture how the millions of hereditary factors contained in the forty-six chromosomes of Man might be sorted out during maturation division, or what possible combinations they could undergo during fertilization, you would shut this book at once. The case is quite complex enough if we study just two different hereditary factors at the same time. This we shall try to do, because we can learn a few important facts about it.

A more complicated hybridization experiment

There are two varieties of guinea pigs. One has a black and short-haired coat, the other a white and long-haired one. If one crosses them, all animals of the first filial generation look alike. They are black and long-haired (Fig. 91). From this we conclude that black and long hair are dominant, white and short hair obviously recessive. This is so far simple and easy to understand. If one now crossbreeds hybrid individuals, segregation of the hereditary factors occurs in the second filial generation. If a sufficiently large number of offspring are obtained, one records on the average a ratio of nine black long-haired animals to three black short-haired to three white long-haired and one white short-haired animal.

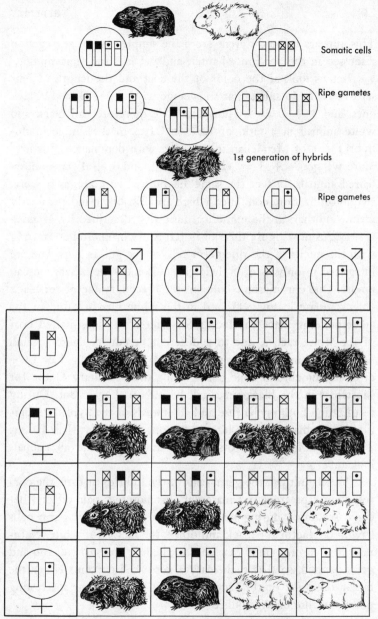

Somatic cells

Ripe gametes

1st generation of hybrids

Ripe gametes

Fig. 91. Crossbreeding of two stocks of guinea pigs demonstrating the inheritance of two different characteristics.

This appears to be different from the simple case of segregation discussed in the previous chapter, and yet it is the same thing.

Let us look at the color of the coat and the length of hair separately. In 16 animals, at an average we got 9 + 3 = 12 black ones, and 3 + 1 = 4 white ones. We therefore had black and white animals at a ratio of 12:4 or 3:1 as in the simple mono-hybrid case of Mendelian inheritance, with dominance. Further-more we got 9 + 3 = 12 long-haired, and 3 + 1 = 4 short-haired animals, in fact the same 3:1 ratio. There are, as it were, two hybridizations combined which are distributed independently of one another, in the numerical ratio first described by Mendel.

Let us now figure this out in terms of chromosomes, starting with the nuclei of the ripe germ cells of the guinea pigs. We are interested in only two of their chromosomes; the others we can ignore. One chromosome contains the hereditary factor for black (black square in Fig. 91), and in the white race we find in the corresponding place the factor for white. The long-haired coat of the white race is due to a different hereditary factor which lies in a second chromosome. We shall indicate long hair by a cross in our chromosome chart, while the hereditary factor for short hair in the black race will be indicated by a dot. During fertilization both nuclei fuse and all animals of the hybrid genera-tion have a black and a white, and a long and a short, hair factor. The two dominant factors prevail and they make all animals look alike.

When, in turn, the germ cells of the hybrids mature, the cor-responding paternal and maternal chromosomes pair off and be-come separated during maturation division. We get germ cells having the factor for black and others the factor for white. Natu-rally, the chromosomes that control the hair coat pair off too and get separated, so that each mature germ cell contains either the factor for short or the factor for long hair. But one interesting point must not be overlooked:

When the corresponding chromosomes pair off, it is a pure matter of chance on which side of the arrangement any given

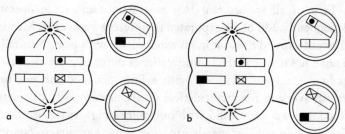

Fig. 92. Formation of different gametes at maturation division.

paternal or maternal chromosome comes to lie. The two pairs of chromosomes that interest us in this case might therefore arrange themselves either in the manner of Fig. 92a or in that of Fig. 92b. Either the paternal or maternal chromosomes lie on the same or on the other side of the center line. Accordingly we get four possible assortments of our hereditary factors in the mature germ cells, each of which has an equal chance to be formed and in a sufficiently large number of germ cells turns up in equal numbers.

Since the same holds for eggs as for sperm cells, four different kinds of egg cells can be fertilized by four different kinds of sperm cells. The possible combinations are most easily examined if set out on a matrix (Fig. 91). On the left-hand side we write down the four possible types of egg cells, across the top of the matrix the four possible types of sperm cells. In the fields of the matrix we enter the results that ensue when the female gamete put down in the horizontal row is crossed with the male gamete of the intersecting vertical row. If we now remember that all the guinea pigs that have the dominant factor for black look black and that all those with the dominant factor for long hair will be long-haired, we have before us the numerical explanation for the special way in which the hereditary factors distribute themselves in this particular hybridization experiment.

Why we discussed this experiment

In this slightly more complicated hybridization experiment two things are of more general interest.

First of all we see that two hereditary factors can operate independently and be incorporated in different germ cells. Mendel had already found that out in his experiments with peas. Naturally the same holds true if the hybrid varieties differ in more than two hereditary factors. Anybody with a little patience and not too great an aversion for mathematical problems can easily calculate, in the same way as we did with our guinea pigs, that with three different pairs of independently operating hereditary factors, the first bybrid generation will produce eight different kinds of germ cells, which will, after fertilization, appear in sixty-four possible combinations of the hereditary factors. If the individuals differ in ten characteristics, under the same conditions we get 2^{10} or 1,024 different kinds of germ cells, combining in a million possible ways.

There are of course limits to this. The number of chromosomes is limited and in some cases very small indeed. In a Drosophila, with only four chromosomes in the mature germ cell, it is impossible that ten hereditary factors could distribute themselves independently by lying on different chromosomes. The numerous factors that are arranged on one and the same chromosome stay together during maturation division, like travelers using the same coach. They are linked and handed on together in this way, provided they do not "get out of the coach" and move into another chromosome. This does in fact occasionally happen, but we need not discuss it in detail.

The following, however, is quite easy to see: the number of linkage groups of factors that show linked inheritance must correspond exactly with the number of chromosome pairs present in any given species. In the course of many hybridization experiments geneticists have found that this is indeed so — new and most elegant proof that the chromosomes are the carriers of hereditary factors.

The extent of the validity of the laws of heredity

To what extent are the laws of Mendel valid? Do all the things we heard about guinea pigs and peas hold true for human beings also?

The answer is that almost all hereditary characteristics so far studied are handed on according to Mendel's laws. But for varying reasons things are not always quite so obvious as in our examples.

One of the reasons for this is that a characteristic feature may be controlled by two hereditary factors, located on two different chromosomes. For instance, the black coat of an animal may be due to two factors, each alone sufficient to bring about black pigmentation. In a white variety there might be two corresponding factors for white. Hybridization of these varieties produces in the first generation black animals only, in the second filial generation fifteen black and one white animal, quite an unexpected ratio, which we can understand at once when we put on our matrix a second black and white factor instead of those for long and short hair. We see at a glance that only once among the sixteen possible combinations do we get conditions that make for the white hair color, while in the other fifteen cases the dominant factor for black is present at least once if not several times.

Another possibility is that each of the factors for black by itself cannot produce a deeply colored black hair coat, but that the final effect is due to a summation of all the factors for black. As a rule not one or two but quite a number of hereditary factors are involved in the creation of any one character. On the other hand one hereditary factor many influence the shaping of a number of characters, and it is one of the most fascinating and at the same time most complicated tasks of geneticists to unravel this interplay of hereditary units.

5. *What Twins Can Teach Us*

Identical and other twins

There are two kinds of twins: those who by their close similarity often cause comment, and those who do not resemble each other more than ordinary siblings. Identical twins have always been considered as something special. We know now that they originate from one single fertilized egg cell, which develops accidentally, not into a single individual but into two, in a way quite similar to the tied-off egg of a newt with which Spemann experimented. Nonidentical twins occur when — exceptionally — two egg cells are liberated into the uterus at the same time and both get fertilized on the way. In this case we get two children who are derived from different germ cells of the same parents and therefore resemble each other more or less closely, as is usual among siblings. Identical twins developing from the same fertilized egg have identical hereditary factors. This is the reason why they are so difficult to distinguish.

This can go so far that even the discerning eye of the mother finds it difficult to tell them apart, and of many a pair of twin girls one has had to wear a red and one a blue hair ribbon to help identification. This similarity can apply to characteristics that only identical twins are known to share. The pattern of lines showing on our finger tips is slightly different in each individual. This is so clearly established that the police accept the identity of fingerprints as definite proof for the identity of the person. But identical twins can have astonishingly similar fingerprint patterns.

The influence of the environment

However, not everything is irrevocably determined by heredity.

The environment in which the inherited formative powers assert themselves is of importance also. This very question — the balance of power between hereditary factors and the environment—made geneticists turn with renewed interest to the study of identical twins.

A thorough study of identical twins revealed quite a few differences traceable only to the influence of environment. Presumably they would be even more frequent and striking if as a rule the external conditions under which twins grow up were not identical also.

More would be known about this if one could experiment freely with human beings, for instance bring up one child in comfort, well nourished and with all the benefits of civilization and careful education, while the other twin was kept short of all this and exposed to negative influences. Only then could one see how identical hereditary material would respond to different environments. But such an experiment would run counter to the rules of human conduct, and thus science is dependent on a few chance observations.

In plants and animals a great number of relevant experiments have been carried out. One can breed hundreds of bean seeds which are as similar in their hereditary factors as identical twins. If one plants them in two different beds, one having good soil, bright light, and a good supply of water, the other one the very opposite, in fact in unsuitable conditions, we get in the first case strong and in the other weak plants, despite their equal hereditary make-up.

Of two litter-mates, belonging to a very uniform race of pigs, one was well fed and the other undernourished. Although growth and deposition of fat are determined by hereditary factors, the change in nutrition turned the two litter-mates, ordinarily very much alike, into opposites — one lean and one plump. This shows that hereditary factors do not shape living creatures as a mold shapes a church bell but that the environment is an important factor in development.

Hereditary factors represent the potentiality to develop characteristics and faculties that have upper and lower limits. Within these limits the environment can influence the properties of plants and animals, as well as of Man. What becomes of a personality is the sum total of the formative influence of hereditary factors and of the environment. However formidable the power of heredity, the environment can inhibit, improve, or enhance not only external features such as growth and shape but every other quality, even the mind and character.

6. *Boy or Girl?*

Identical twins are not only similar in their external features and in their mental make-up, they are always of the same sex, either both male or both female. Nonidentical twins can, like any other siblings, be of different sex. We know that identical twins have identical hereditary factors that are contained in the chromosomes of the nucleus. Is it possible that sex too is determined by chromosomes?

A little worm as chief witness in a fundamental case

For the solution of this question, as for many other problems in biology, experiments on lower animals, such as bugs or little worms, have been of decisive importance. Thus a little insignificant parasitic worm, with the pompous name of *Ancyracanthus cysticicola*, surprised biologists by the fact that the female had in all the nuclei of its body cells twelve, while the male had only eleven chromosomes.

In the maturing egg cell (left-hand side of Fig. 93) six pairs of chromosomes are formed which get separated during maturation of the egg, so that every ripe egg contains six chromosomes.

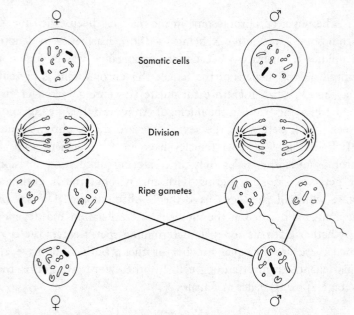

♀   ♂

Somatic cells

Division

Ripe gametes

♀   ♂

Fig. 93. Sex determination in the parasitic worm, Ancyracanthus.

That in this animal they are not threads but short stumpy rods is, in this context, of no importance.

In the maturing sperm cell (right-hand side of Fig. 93) the eleventh chromosome has not got a partner; it stands all alone in the middle of the spindle, and when the pairs as usual separate, it has to join the one or the other side. Thus we get two kinds of sperm cells, the ones with six and the others with five chromosomes, both kinds in equal numbers.

The nucleus of the egg invariably contains six chromosomes. If a sperm cell with six chromosomes fuses with the egg we get after fertilization 6 + 6 = 12 chromosomes (left bottom corner of figure). Since all body cells derive from the fertilized egg, they too contain all twelve chromosomes. We already know, however, that this is a distinguishing feature of the female worm! If an egg gets fertilized by a sperm with five chromosomes (bottom right-hand side) we get eleven chromosomes and the animal is a male.

The unpaired chromosome in the male has been called the X chromosome — the letter X being used in mathematics to denote an unknown quantity — because its significance was at first not realized. In the nucleus of the female this chromosome is present as a pair. We can therefore formulate this strange arrangement in the following way: In the nuclei of Ancyracanthus are chromosomes which determine the sex. These are the X chromosomes (black in Fig. 93). The females have in each nucleus two X chromosomes, the males only one. During maturation division we get two kinds of sperms, with and without the X chromosome, while all ripe eggs have the X chromosome. The sex of the animal is decided at the moment of fertilization and depends on whether a female- or male-determining sperm enters the egg. Since there are an equal number of them, both kinds have an equal chance to meet an egg cell. In other words we get on the average as many males as females.

X and Y

The reader will now ask at once whether sex determination as we observed it in our little worm is the general rule and whether the universal phenomenon of the presence of an approximately equal number of males and females is in this way simply and comprehensively explained.

The answer, for most animals and a great many plants, is yes, and this is one of the most extraordinary discoveries of genetics. There are certain deviations concerning minor details, the most common being that the male has also two sex chromosomes, which, however, are functionally different and also show external differences in size and often in shape. They are called the Y and the X chromosomes (Fig. 94). In this case all ripe egg cells contain an X chromosome (left-hand side of Fig. 94); the male produces two sperm cells, one with the X and one with the Y chromosome (right side of Fig. 94). During fertilization by the former we get XX nuclei, therefore females; and when by the latter, XY nuclei, therefore males. This type of sex determination holds for Man too. In Man the conditions are less favorable for investigation than in the fruit fly and many other animals, because of the smallness of the chromosomes and their great number. It required most careful studies and much time and effort to elucidate the facts.

On the basis of these findings it is to be expected that on the average an equal number of boys and girls are born. In Man we have actually a preponderance of male births, and in some animals we get a considerable deviation from the expected 1:1 ratio. The explanation lies mostly either in an unequal mortality rate of the embryos in both sexes during their early stages of development or in the fact that of two types of sperm cells, one may be at a disadvantage in fertilization through lack of viability or other causes, for instance diminished motility. If the male-determining

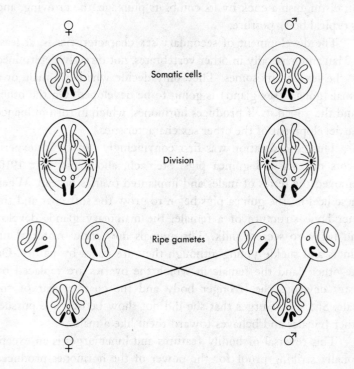

Fig. 94. Sex determination in Drosophila; the sex chromosomes are in black.

sperm cells are more agile, they will on the average reach their goal sooner and we shall get more males than females.

The secondary sex characteristics

When we talk of sex determination we mean by this principally the decision as to what kind of gonad is going to develop in an individual. An animal with the male gonad, the testes, is a male; an animal with the female gonad, the ovary, is a female. The externally noticeable differences between the sexes concern other characteristics and have as such primarily nothing to do with the development of the germ cells. A man differs from a woman by the growth of a beard, by his deep voice, by a more strongly built body, just as a male red deer has antlers and as we can distinguish a cock by its comb, its plumage, the crowing, and its typical body posture.

The development of secondary sex characteristics is, at least in Man and generally in other vertebrates, not directly determined by the sex chromosomes. They only decide whether a male or a female gonad (sex gland) is going to be developed. On the other hand the gonad itself produces hormones, which in turn influence the development of the other sex characteristics.

This co-ordination was first convincingly proved by experiments on rats and guinea pigs. Steinach, about the year 1910, removed the testes of males and implanted ovaries instead. When these heal in, the guinea pigs begin to grow the soft coat and the finer bone structure of a female; the mammary glands develop and begin to secrete milk. The animals now show maternal instincts and suckle young although they are males by nature. On the other hand the female in which the ovaries are replaced by testes develops the stronger body and the coarser coat of the male. She shows urges that she did not show before; she pursues other females and behaves toward them like a male.

This reversal of bodily features and inner urges is an exceptionally striking proof for the power of the hormones produced in the gonads. In vertebrates even the mere removal of the gonads, without implantation of those of the other sex, has very striking consequences. The secondary sex characteristics regress more and more, or, if the operation is performed at an early age, never develop. This has been known for a long time and has been used by Man for his own ends. A castrated cock loses its fighting spirit and turns into a fat capon, a defiant bull turns into a docile ox and a willing beast of burden. Man has never been known to have any inhibitions about interfering ruthlessly with the mechanisms of Nature, if this offers an advantage.

VII

THE EVOLUTION OF SPECIES DURING THE EARTH'S HISTORY

---

1. *About the Transmutation of Species and the Origin of Life — a Chapter with an Unsatisfactory Ending*

A professional collector of beetles is not content to kill the six-legged objects of his studies and to pin them in orderly rows into his collector's cases, but will try to classify them and will not rest until he has found out the exact name of each one of them. This is not a simple matter, for more than 300,000 beetles have been described so far. That it is possible at all to find one's way about within the multiplicity of forms, the collector is indebted to the great Swedish naturalist Carolus Linnaeus or Linné (1707-1778). It was he who established the first useful system of classification, not only for beetles and other insects but for the whole of the plant and animal kingdoms. With his penetrating powers of observation he discovered the characteristics one has to look for in order to establish the species to which any given animal or plant belongs. He did not worry, however, about the origin of the species under observation, his point of view being that there are as many species as God created in the beginning. Biologists after him were not so easily satisfied. They had noticed that animal and plant species are not immutable and rigidly set off against one another.

Within a species there exist smaller subgroups, the varieties, which differ from other varieties by a certain number of heritable factors. The border line between varieties and species, and between species and species, is often indistinct and remains to this day a matter of personal opinion. Besides, one has learned to pay attention to witnesses from the earth's past, which testify very definitely against the view of the immutability of species.

Witnesses for life on earth in past millennia? Not a very reliable source, one would think, considering that witnesses contradict each other about events that happened only a few weeks ago. But there is a history in stone, reaching from the early periods to the present day, and it is more reliable than the memory of Man.

How fossils are formed

In the year A.D. 79 the town of Pompeii was destroyed by a violent eruption of Vesuvius. The majority of its inhabitants could save themselves, but many who sought protection in their houses, or went on too long trying to save their possessions, perished and were buried in layers of ashes many feet high. Later the ashes were changed into a mass of mud by torrents of rain and the corpses were enclosed in the hardening mudlike casts as in a mold. As the corpses decayed in time, the hollow molds remained. When, in the course of nineteenth-century excavations, such molds were found, the Italian archaeologist Fiorelli had the idea of filling them with plaster of Paris. In the museums of Pompeii and Naples a whole collection of such plaster-of-Paris casts is on show and one sees the unfortunate victims in exactly the positions in which they had been overcome, in the clothes they were wearing, and some with well-preserved facial expressions.

In a similar way we get to know the shapes of animals of far earlier times, and long extinct. Whether they perished during volcanic eruptions or, as happened more frequently, just sank into the muddy deposits of water, they produced by their decay hollow shapes, which have become filled up with casts of slowly hardening mineral substances. Now these stone casts show us the exact shapes of the former organisms.

There are other ways in which fossils can be formed. When rivers run thick with yellowish clay, deposits of the finest clay are formed in quiet backwaters. This clay is so fine that the corpses of small animals leave the most wonderful imprints, even of their most delicate parts. If as a result of a flood a layer of different material is deposited on top of this the imprints get filled in and slowly such deposits form hard rock, preserving the imprints. On the other hand the organic remnants of animals and plants can be impregnated with mineral substances from the surroundings, so that they are turned into stone in the truest sense of the word. The prospects are most favorable for animals with hard skeletons, which naturally leave better imprints than soft parts and are altogether less perishable.

Since time immemorial water has transported solid particles of earth and rock, which, settling in the beds of sluggish rivers, on the bottoms of lakes and the sea, formed vast deposits during the millennia. Repeatedly whole land masses became submerged or the sea bed was raised and turned into land. The crust of the earth formed folds and wrinkles and in this way produced mountains. This explains why even in the highest mountains we now find old marine deposits with their enclosed remains of marine animals.

The theory of evolution

Fossils are direct evidence of the fact that living organisms have changed with time. In the oldest deposits on earth that contain animal remnants, traces of vertebrates are completely absent. In the next younger deposit we find fossilized fish of a very primitive type, and in still younger deposits we find remnants of fish resembling more closely those now living. Then come layers or strata containing remnants of amphibia and reptiles besides those of fish. Some of the reptiles are monstrous giants, and their skeletons are among the most highly valued exhibits in museums of natural history. During the chalk, or so-called Cretaceous, period most of these extraordinary animals became extinct and were replaced by the most highly organized vertebrates, the birds, the mammals, and finally Man.

If fossils of all forms that ever lived existed, they would demonstrate to us every step in the evolution of the various species. As most organisms decay completely, leaving lasting traces only under the most favorable conditions, the compilation of a complete history of the evolution of animals and plants will remain an elusive mirage. We must consider ourselves lucky if we can discover occasional traces of the evolutionary history of one or the other group. Thus it was possible to study from the fossilized remnants of horses how the original five-toed leg turned into the foot of the contemporary horse by a gradual stronger development of the middle toe and the disappearance of the side toes (Fig. 95). The middle toe now touches the ground with the tip of its hoof only (which is nothing but a very strong toenail)— an excellent prerequisite for fleet-footedness. A human runner tries to imitate this technique but Nature has not equipped him so well.

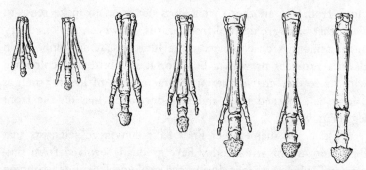

Fig. 95. Anterior foot (with several toes) of the ancestor of the horse and its evolution into the hoof of the modern horse.

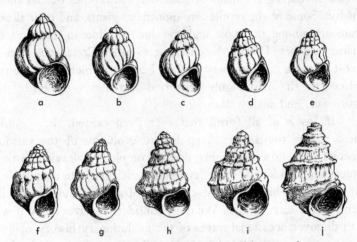

Fig. 96. Evolution of one species of snail into another.

The illustration (Fig. 96) is an example of such an evolutionary series in less highly developed animals. In the deposits of a fresh-water lake in Yugoslavia layers of different ages were found to contain different species of a water snail (a-j). Intermediate layers contained intermediate stages, bearing witness to a gradual change from one species into another.

Certain groups of animals that we consider anatomically closely related, as, for instance, lizards and birds, differ now so much from one another that it does not seem to make sense to claim that birds are derived from lizards. However, the discovery of the famous Archaeopteryx in the Jurassic slate of Solnhofen in Bavaria provides us with a link, for it is a bird that combined with its feather coat numerous characteristics of lizards, such as a long tail, toothed jaws, and well-developed toes on the front legs (Fig. 97).

These and many other finds have convinced scientists that the living organisms of today have gradually evolved from simpler and simplest shapes during the geological periods, spanning a time for which our human brain lacks the power of comprehension. This forms the basis for the theory of evolution. How the transmutation of species took place, especially in lower animals, is not clear at all. Their origin reaches back into times from which no fossil traces of life have come to us. Of course numerous primitive organisms continue to exist, such as sponges and the unicellular plants and animals. They too must have changed since those early days, and in their present shape they cannot be regarded as the ancestors of more highly organized contemporary organisms any more than the now living apes can be considered the ancestors of Man. What they have in common are the roots only. It is not possible to compile a family tree for organisms now living, naming parents, grandparents, and great-grandparents.

Fig. 97. Fossil of a primitive bird, Archaeopteryx. Note the reptilelike head and primitive wings.

All that is living now represents a cross section through the different branches of the ancestral tree. The true ancestors of our animals and plants have in the best of cases come to us as fossils, and although we have only an incomplete knowledge of them no biologist nowadays doubts the fact of evolution.

Fossils, however, are not the only evidence for evolution. A number of facts related to the design and development of animals and of Man would be quite incomprehensible otherwise. The appearance of gill slits in the human embryo can be understood only if we assume that our early ancestors were aquatic animals, and that an old blueprint was used as a basis for the new one. The existence of a great number of reduced muscles in the pinna of our ear which at present serve no functional purpose does not make sense, except as a heritage from ancestors for whom movable pinnae were of biological importance. If one wanted to discuss all the features of Man that testify to his past it would fill a whole book. Not only Man and the vertebrates but all great phyla of animals show a uniformity in their basic design that can be explained simply and convincingly by a common origin.

Spontaneous generation

If we assume that the higher forms of life evolved from simpler ones our trend of thought leads us back to the simplest living organisms and to the roots of life itself. Life must have originated on this earth at some point in time. When our planet was still a hot glowing mass like the sun, no life could have existed on it.

The earlier naturalists saw no problem in this because they firmly believed in spontaneous generation, which means the creation of living things from dead matter. Aristotle relates that eels develop from worms and worms from mud. A report from the sixteenth century tells us that young mice develop from flour and from a dirty shirt; the parent mice that had their nest in the shirt seem to have escaped the attention of the observer. With the invention of the microscope and the discovery of the microcosmos the study of spontaneous generation got new impetus. Could one not observe in a glass of water the development of masses of small animals and plants? Then came Pasteur, who destroyed these fantastic ideas. Not a trace of life develops in a glass of water if by thorough boiling one kills any germs of life in it and if one takes care that none enter into it from the air. There are innumerable germs of lower organisms that are dispersed by the wind with the dust from dried-up puddles. They come to life again as soon as they find suitable conditions. We are now convinced that the protozoa, which are so small that only the microscope conjured them up from the world of the invisible, are not as simple as they might appear to the naïve observer. Into the nucleus of an amoeba are packed basically the same mysteries of heredity as into the nucleus of a human egg cell. The bacteria, the lowest of all plant forms, are simpler than amoebae in their general organization and so far as their nucleus is concerned. Their mechanism of reproduction, their flagella, by means of which some of them move about, and their adaptability to a wide range of environmental conditions are so remarkable that no modern scientist would maintain that such organisms could originate from inorganic material from one day to the next.

Now our earth is old, and very likely conditions no longer exist under which living material could arise from nonliving matter. About three thousand million or so years ago,* when the earth was in an adolescent stage, very different things happened in its hot gas envelope. The first atmosphere was free of oxygen and the gases it contained were hydrogen, water vapor, simple carbon-hydrogen compounds, and ammonia. One assumes that chemical reactions in that atmosphere were encouraged by violent electric discharges. The chemist Stanley Miller in Chicago carried out a very interesting experiment. He exposed in the laboratory and artificial atmosphere of methane, hydrogen, water vapor, and ammonia for eight days to heat and electric discharges, and he could demonstrate the formation of a certain amount of amino acids. These are nitrogen-containing compounds that are the simplest components of the proteins, the basic substances of living matter. A synthesis of organic compounds from inorganic substances was thus successfully achieved under conditions that may have existed in the gaseous envelope of our earth before the appearance of life. Admittedly it is a long stretch from those first amino acids to protein molecules and to living units. One would like to see the intermediate stages, and for a while it was believed that they had been found.

Borrowed life

When it was recognized that microscopically small protozoa and

---

*The age of the earth is variously estimated. Some estimates put it at more than four billion years.

bacteria are the cause of certain infectious diseases, an intensive search for other pathogenic germs began. As the tiny criminals that cause malaria and Asiatic cholera had been found, it was now hoped that those that cause all the other epidemics among non-resistant organisms might be found. It was important to discover them in order to be able to fight them more efficiently. It is difficult to fight an unknown enemy.

A large field of work opened up for the inquisitive scientists. A whole range of infectious diseases besets human society from the plague down to the common cold. Their number becomes legion if we add those that befall animals and plants. If domestic animals and economically important plants are stricken they are not the only ones that suffer; the money bag of farmers and planters is also affected. Quite often such epidemics are followed by poverty and famine. All this is reason enough to employ all means at our disposal to disentangle cause and effect.

Great success was achieved in a limited field. We got to know the cause of typhus and cholera and the importance of the tubercle bacillus, but so far we have looked in vain for the instigator of the common cold, although this is without doubt infectious. The minute bacteria that cause the plague and the bacilli that cause the usually fatal swine fever in pigs have been discovered. On the other hand, not even the best microscope has disclosed the cause of measles or of poliomyelitis, of the dreadful hydrophobia or of the much-dreaded foot-and-mouth disease of cattle. And yet there must be germs in the infected animals' body fluids, because the disease can be spread by injecting these into healthy individuals.

Tobacco plants have an infectious disease called mosaic disease because it makes the leaves go spotted. If the juice from diseased leaves is passed through special filters with such small pores that even the smallest bacteria cannot pass through, the filtrate, although free now of any bacteria, will still infect healthy tobacco plants. This mysterious something we call now-adays a filterable virus. Virus is the Latin for poison. The particles

Fig. 98. Viruses as viewed through the electron microscope. Left, virus of the tobacco mosaic (enlarged 10,000 times). Middle, a virus of the potato (enlarged 30,000 times). Right, virus (bacteriophage, enlarged almost 30,000 times).

of this infectious but unknown kind of poisonous substance must be smaller than even the smallest bacteria.

From guessing and believing to understanding and knowing is a far cry. A lot is known nowadays not only of such unimaginable small bodies as those associated with the mosaic disease of tobacco plants, but of similar agents causing foot-and-mouth disease, hydrophobia, measles, smallpox, and the common cold, poliomyelitis and many others, originally ascribed to the action of bacteria. The term "virus diseases" has come to stay, and volumes have been written about them. The more that was found out, the more fascinating it all became.

What does a virus actually look like? This seems to be a pretty unanswerable question, since not even the best light microscope shows them up. The first picture of a virus was, however, revealed by the modern electron microscope, which can even photograph such small units of matter as large molecules (Fig. 98). The viruses look like bacteria on a much smaller scale and like them they are, according to their kind, spherical, rod-, or thread-shaped, or they may be tailed spheres, looking like miniature tadpoles (Fig. 98).

Exciting as this may be, the really thrilling things are the other details that were found out about the nature of viruses. First of all, they are not all equally small. Even among them there are giants, like the psittacosis virus in parrots, which almost reaches the size of the smallest known bacterium. In other cases the virus is as small as a single protein molecule, for example the virus responsible for foot-and-mouth disease. Stanley, the American Nobel scientist, succeeded in isolating the tobacco mosaic virus in its pure form. It is a protein of the same kind as is found in the nuclei of animal and plant cells (desoxyribonucleic acid). It is known that the hereditary material in the chromosomes consists of protein molecules of this kind. It was recognized that the single virus particle in a diseased tobacco plant is nothing but a large protein molecule of this type. And — still more extraordinary — the chromosomes of the nucleus are capable of replication by growth, in which one protein molecule produces a second similar one, which is separated from the other during nuclear division and thus becomes the carrier of heredity (p. 234). The virus particles can replicate themselves and reproduce their kind, whereby they make devastating use of the substance of their host, be it a tobacco plant or any other virus-diseased organism. Besides, Stanley could show that the virus particles, in other words the protein molecules, which he extracted from the mosaic tobacco plants, will, after being injected into healthy plants, multiply vigorously and cause the disease.

Viruses that cause diseases in Man and animals are often structurally and chemically much more complicated than plant viruses. As the largest among them look superficially like the smallest bacteria, it appeared as if with the viruses there had been discovered forms of subbacterial life. But are viruses really living things?

Naturally, attempts were made to grow them on artificial culture media, as had been done so successfully with disease-causing bacteria. This failed: it is impossible to grow viruses by bacteriological methods. They can be cultivated artificially only in living cells. This can be done in different ways. Some viruses are injected into chicken eggs, others are let loose on bacteria. Thus the largest and the smallest cells have been recruited for this enterprise. If the virus is to reproduce, it has to be in a living cell. Viruses cannot grow on dead organic material.

In this way it was found out that viruses lack one feature that is characteristic of life: they have no metabolism and therefore no life of their own. They make use of the metabolism of the foreign cell that accommodates them and use its substance to build up new virus particles, until the host cell perishes and bursts. The viruses are then liberated and remain inactive until they reach other cells from which they may borrow new life for themselves. With the discovery of the viruses it was thought that we had arrived at the root of all life. But fierce discussions ensued as to whether viruses are to be considered alive or whether they are the forerunners of living matter. On sober reflection the view was reached that these structures, though of simpler organization than bacteria, cannot be considered forerunners of living matter, since they need the environment of living cells for their continued propagation. It is conceivable that conditions on this earth were once so different that at the time of their origin the ancestors of the viruses obtained the material and energy for their reproduction not from living cells but from dead inorganic matter. In this sense viruslike units may have played a role in the origin of life early in the earth's history. However, for the time being such ideas are mere fantasies and it is not profitable to elaborate on them.

Occasionally one hears the opinion that life on this planet could never have been created from nonliving matter, but that the germs of life have reached us through shooting stars or other celestial vehicles that came to us from outer space. This only shifts the problem from one star to another.

2. *Darwin's Thoughts on Natural Selection*

To some people Darwinism means that Man is supposed to be descended from the ape. This is quite erroneous for several reasons. First of all, nobody seriously claims that the now living apes are the ancestors of Man: the theory of evolution only postulates that both have a common ancestor. Furthermore, Darwin was not

Fig. 99. Varieties of the pigeon.

the first to speak of evolution. This was done for instance by Lamarck before him. It was Charles Darwin who, about a hundred years ago, collected abundant and convincing material evidence that consequently led to the general recognition of the theory of evolution. Finally, it was not the recognition of the change and gradual development of species that was Darwin's chief contribution but his attempt to explain it as a process due to natural selection. This was his new and ingenious idea. When biologists talk about "Darwinism" they refer to the much-disputed process of natural selection, not to evolution as such, which is so well documented that hardly anybody doubts it.

Artificial selection

Darwin's attempt at an explanation of the origin of species was based in part on his experiences as a breeder of animals. The modern varieties of the domestic pigeon are the result of many thousands of years of breeding experiments. They all are descended from the wild rock dove (Fig. 99) from which by now they differ considerably in appearance and habits: in the color of their plumage, in the development of ornamentally feathered feet, in their special voices (trumpeters and laughers), or by their habit of somersaulting during flight (tumblers) or having showy hoods or fantails. The fanciers love such peculiarities, and the breeders try to create them.

According to Darwin, three factors have to work together to achieve this: first of all the fact of natural variability, which means that offspring of one pair of parents differ, if only in small details; second, the inheritance of parental characteristics, which is not limited to the fact that eggs of pigeons produce pigeons, but which also applies to special peculiarities of the parents; third, the deliberate choice of the breeder, or artificial selection. For example, if he wants a variety of pigeons with a large tail he makes use of the natural variability of the tail and chooses for breeding only such animals as have more than the usual twelve tail feathers. These will hand on their special characteristic to their progeny, and through continued selection along this line one eventually arrives at fantail pigeons, which have forty-eight tail feathers. Other races of pigeons have been produced in a similar way by the selection and intensive breeding of the desired characteristics.

Natural selection

Variability is found in wild species too, and obviously the hereditary laws are the same for domesticated and wild animals and plants. But who decides in Nature what is going to be bred? It was one of the greatest biological insights of all times to draw conclusions from the deliberate choice of the breeder about the way Nature makes her choice.

What did Darwin mean by natural selection, the focal point of his work? Let us assume a female house fly lays 100 eggs. From them hatch 100 larvae and we nurse them carefully until they have grown into 100 buzzing flies, which means about 50 pairs. A male and a female, under the best conditions, produce another 100 eggs before they die. Now we have 5,000 flies, or 2,500 pairs. Under suitable conditions we can get a succession of 15 generations a year. If we had the means and helpers to nurse all of them carefully, so that none perished, all the continents of this earth would, within nine months, be covered by a continuous layer of our charges and before a year was out the tops of church towers only would emerge from a sea of flies. If you do not believe this take paper and pencil and calculate. A house fly is a quarter of an inch long, an eighth of an inch broad and high. The continents of this earth have a surface area of about 90,000,000 square miles.

In reality we are spared such a flood of house flies. Their number on this earth is now not significantly different from what it was when we were young. The reason is that of about 100 offspring of a pair of flies 98 perish before they have a chance to reproduce. This is the only way to keep the number of individuals at a constant level. The majority of the progeny is destroyed in the daily struggle for survival. Numerous are the natural enemies that live on all the developmental stages of the fly. Besides, there are environmental vicissitudes to be endured, like downpours of rain, bitter cold, chemical sprays like DDT, and others. Lastly, there is the struggle against the other members of the species that compete for subsistence, which is all the fiercer if the number of individuals temporarily increases through favorable circumstances.

If we now remind ourselves of the variability of hereditary characteristics it will be clear that animals have the best chance to survive in the struggle for existence if they deviate, if ever so slightly, in an advantageous direction, while a less advantageous deviation hastens their death. Just as the animal breeder chooses for breeding those animals that show desired features at their best, Nature lets those forms survive and reproduce that are best adapted to the prevailing environmental conditions. This helps us to envisage a gradual perfection of living beings and their adaptation to the environment, even without the intervention of a purposefully planning creator.

You may say that not all animals multiply as profusely as house flies. But these are only differences in degree; in fact all living beings show considerable overproduction of progeny. Even with elephants and their few young one could get astonishing numerical results if they could be spared the struggle for existence. A single pair with a life span of 100 years and a progeny of 6 elephants per elephant pair would produce a herd of 19,000 animals within 750 years. Since their number is not increasing they too must suffer extraordinarily high losses and natural selection must be correspondingly fierce.

There is little doubt that natural selection takes place, and this has been repeatedly observed. We have already discussed an experiment that showed that fish well adapted to their background evade their pursuers more easily than the less well-adapted ones. A spider is easily discovered and eaten by the robin if all its limbs are showing, but it is very difficult to discover if it assumes its protective posture. Catastrophic damage was done to coniferous woods by certain caterpillars, and only a few firs survived, because their needles had a higher turpentine content than the average. After an extremely severe winter there were found innumerable dead frogs, which had not dug themselves in deeply enough for hibernation. Only the frogs with a well-developed instinct urging them to dig deeply into the mud of their pond survived at that time.

In these as in other cases selection works in the direction of a heightening of the useful adaptation provided that the advantageous trend, be it an instinct, or a chemical property, or any other characteristic, is inherited by the progeny.

Whether natural selection alone can explain the origin of species, with their abundance of often most intricate adaptive devices, is a different question. Darwin himself did not assume this. It was left to his successors to use such proud words as "the omnipotence of natural selection."

Opinions differ about the extent of the powers of natural selection. Many think that it is too much to believe that the mere favoring of small deviations brings about the wonderful designs in form and function shown by organisms. It must not be overlooked, however, that natural selection has one mighty ally: namely, time.

Nature is in no hurry

Until recently all data about the age of rocks and the time taken by the species to evolve in were based on estimates that deviated considerably from one another, as there were any number of sources of error. Now, however unlikely it may sound, one can make rather accurate pronouncements in this respect. This is due to the discovery of the radioactive substances and their decay.

By a strictly regular decay, the speed of which is quite independent of environmental conditions, the element uranium, after passing through a number of intermediate stages, produces lead as end product. If the uranium and lead content of rock is analyzed, the proportion of these two substances indicates how long this decay has been going on; in other words, we know the age of the rock. The objection that uranium might have turned more slowly or more quickly into lead in earlier days can be ruled out with confidence, because the process of decay leaves certain traces in the rock from which one can deduce its speed. It has always been the same.

It has been calculated that the oldest deposits in the sea are 1,500 million years old. The oceans must be older than this, for the deposits to have been laid down in them.

The oldest strata rich in well-preserved fossils are those of the Cambrian. At that time there already existed highly organized animals, such as starfishes and crustaceans and so on, but only the most primitive forms of vertebrates. Since then the phylum of the vertebrates has had 600,000,000 years' time to evolve in.

---

3. Darwin's Theory in the Light of Modern Genetics

Darwin had assumed that the hereditary characteristics of animals and plants are variable and that the variations are heritable. Only half a century later geneticists checked up thoroughly on whether these premises of Darwin's theory of natural selection were valid. For a time it looked as if this meant the end of Darwinism. But Darwin prevailed.

Let us take a closer look at the variability of hereditary characteristics

If one wants to study variability carefully it is best to choose a trait that can be measured accurately. The size of bean seeds, for example, can be easily measured. It is in fact a hereditary characteristic, because there are varieties of bean plants that produce large seeds and others that produce small seeds. We have already mentioned that the development of hereditary characteristics depends not only on the make-up of the gene but on environmental conditions as well. This can be observed very clearly in our example. Beans belong to those plants that can be continuously reproduced by self-fertilization. Thus one can prevent hybridization with other varieties and obtain material of completely uniform characteristics.

The progeny produced by selfing individuals with uniform hereditary characteristics are called true breeders. Yet despite this uniformity each of the plants produces seeds of unequal size. If the length of all the beans produced is measured, it will be found that one size is the most common, say seeds five eighths of an inch long. Besides, one gets larger and smaller beans in decreasing numbers. This becomes quite evident if we arrange the seeds of a plant in separate compartments according to size. If the highest points of the different compartments are connected with one another a typical distribution curve is obtained (Fig. 100), which shows us at a glance the correlation between the size of the beans and their numbers. In the following discussion we shall just draw the curve and imagine compartments with beans under it.

The kind of variability just described, which yields the distribution curve (Fig. 100), signifies that a size of five eighths of

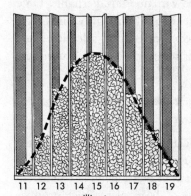

11 12 13 14 15 16 17 18 19
millimeters

Fig. 100. Normal distribution curve of variation in length in the bean.

an inch is the heritable characteristic as it appears under normal circumstances. All seeds of the plant, as it were, try to reach this size. The environmental conditions, however, have an important say in the matter. Some seeds develop on a sunny, others on a strongly shaded, part of the plant, some on a strong, others on a weak branch, some are favorably, others unfavorably joined to the nourishing flow of sap. Thus the hereditary factor for size can be strengthened or inhibited for all kinds of reasons. As a rule the stimulating and the inhibiting conditions will balance equally; occasionally one will have more influence than the other. Large aberrations from the normal median size are therefore relatively rare. In other varieties of beans, seed length will have a different average value (for example, three eighths of an inch).

Artificial selection that ends in failure

If we use the largest and the smallest seeds of a true-breeding bean plant, we get, under identical conditions, plants that produce seeds similar to those of the mother plant, because the factors for size are the same in both. This result cannot be altered, even if for generations one uses always the largest seeds for breeding. Hence, selection from true breeders proved ineffective. This was discovered several decades ago by the Danish botanist and geneticist Wilhelm Ludwig Johannsen, much to the surprise of his fellow scientists. It had not been found before, because uniform animal or plant material had not been used in breeding but always a mixture of varieties.

If seeds are collected from a whole field of beans grown from a number of varieties with differing hereditary characteristics — for simplicity's sake let us assume three — the harvest will again consist of middle-sized, large, and small beans. However, the range of sizes will be wider, because the effect of environmental conditions is added to that of heredity. The curves representing the variability within each of the three races overlap. The actual number of beans of different sizes is obtained by summation of the three curves. As a result we obtain the bold curve shown in the figure, and it is not immediately evident that it represents the combination of three varieties. Selection from these seeds will be successful. If especially large seeds are chosen, there is a chance that it is their hereditary factor that made them larger, and the average size of bean seeds can be raised accordingly from half an inch to nearly five eighths of an inch.

Of course there are limits to successful outbreeding. When one has picked out the seeds with the factor for the largest size, further selection fails to produce an increase in size. Johannsen therefore concluded that artificial selection can do nothing but disentangle a mixture of variations and that a continual elaboration of a character as it was postulated by Darwin is impossible. Consequently Johannsen, and many other geneticists with him, believed that this was the end of the theory of natural selection.

Mutations and their significance

Such reasoning, however, is correct only if hereditary characteristics are once and for all immutable. For quite some time odd cases had become known in which hereditary factors suddenly changed. They are now called mutations, from the Latin *mutare* — to change. Thus within a herd of sheep, which so far had produced nothing but normal straight-legged members, normal parents might one day produce a bowlegged lamb, the progeny of which would show inherited bowleggedness. This variety was for a time greatly favored by farmers, because such sheep could not jump over low fences. The German sausage dog or dachshund very likely came about in a similar way. In another instance a hornless variety of cattle originated just as suddenly and unexpectedly. Sometimes one finds among flies with well-developed wings some specimens that have a pair of short stumps only. If one breeds from them, this special characteristic proves to be heritable. All such sudden changes are brought about by mutations of genes and are therefore heritable.

The importance of mutations as a source for new species may seem to be negligible. But certain adaptations found through them a simple and immediately convincing explanation, for example the frequent occurrence of stunted-winged or wingless insects on remote islands. One of the best examples is Kerguelen, an isolated windswept island. All indigenous insects such as the nine species of flies living there, even the butterflies and beetles, are incapable of flight. On the continent this would be of great disadvantage; on this island underdeveloped wings are an asset. Those insects that rise into the air are in these stormy regions all too easily carried out over the open sea, where they perish. Among artificially bred flies the sudden appearance of forms with heritably stunted wings has been repeatedly observed. It is reasonable to assume that the flies of Kerguelen have their origin in corresponding mutations, which on account of the peculiar living conditions on the island superseded the original winged forms in the struggle for existence.

Although mutations might explain such isolated instances, they seemed of too rare occurrence to explain the innumerable and complicated adaptations necessary for the evolution of different species. According to Darwin an abundance of hereditary variations is necessary from which in the struggle for existence the most suitable ones are selected and promoted. It was a second turning point in the history of Darwinism when geneticists found that in fact mutations occur very frequently.

Striking aberrations like bow legs or stunted wings are of course rare. The new data came to light after careful and laborious studies. During several decades the American geneticist Morgan and his research workers bred millions of little fruit flies and studied them in the greatest detail. They found that the obvious mutations, like white-eyed flies among normal red-eyed ones, or the already mentioned animals with vestigial (stunted) wings, were relatively rare, but that on the other hand small but nevertheless heritable mutations were very frequent, differing only in degree. They might involve any organ or function of the body, be it the color of the eyes or the wing pattern, the number and distribution of bristles on the body, the segmentation of the body, its color and patterning and many other characteristics. Within a short time Morgan's breeding experiments presented us with hundreds of new mutations, and of course hundreds of thousands may have occurred unnoticed in Nature within the same brief span of time. Nature works with an abundance of material compared to laboratory experiments.

When the German botanist Baur started to cultivate snap-dragons his experiences were similar, but with a significant difference: he found that mutations occurred far more frequently than in the fruit fly. Even within most carefully selected true-breeding strains of this plant, his discerning eye discovered on the average among ten plants at least one that showed a heritable mutation. This might relate to the color of the flower, the way of branching, the earlier or later time of flowering, the color of the leaves, and many other characteristics. Many of these mutations do not affect the life of the plant and many are even disadvantageous. On the other hand, many are of a kind that might under certain conditions be of advantage in the struggle for existence. There are certainly many animals and plants the hereditary characteristics of which are less liable to mutate than those of the fruit fly or the snapdragon. But it is just as certain that countless mutations will be discovered in all kinds of organisms if looked for with the assiduity of a Morgan, a Baur and a Muller.

Yet another important discovery, which strongly supports Darwin's theory, was made in modern genetics. It was found that the rate of mutations can be increased artificially by external influence. If, for example, fruit flies are exposed to X rays, as they were by the American Nobel Prize winner, Hermann J. Muller, or bred at an exceptionally high temperature, the germ cells so maltreated show greater numbers of heritable mutations. Now it is quite possible that during the earlier periods of the earth's history, high temperatures or other extraordinary external conditions favored the production of mutations, and that this high variability of organisms led to the formation of new species and natural selection.

The appearance of a great number of mutations is welcome so long as the less favored ones are eradicated by ruthless natural selection, and only the more viable ones preserved.

Nowadays we have entered an age of developments that is uniquely threatening. Human health is endangered by exposure to large amounts of Man-made radiation, without the balancing safeguard of natural selection.

The first warning signs appeared some time ago. The early pioneers working with X rays were shocked to find that they as well as their patients began to suffer from the effects of this radiation. It happened quite unexpectedly and often a long time after exposure.

Nowadays radiation is produced not only for use in laboratories, for X-raying broken limbs, for healing growths, or for experiments on the induction of great mutability in organisms. We now split atoms, a process that releases not only an undreamed-of amount of energy but at the same time radioactivity, the effect of which is considerably stronger than that of X rays. It is horrifying to realize that people who were within the active radius of the first atom bomb dropped over Hiroshima are at this very moment still dying a slow death.

These ruinous if delayed effects on our health are the dreadful price we pay for the achievements of modern technology. But still more sinister because more far-reaching is the harm done by radiation to the germ cells, the carriers of heredity. This concerns the future of mankind.

Whenever an atom bomb is exploded, the atmosphere is likely to be polluted with a great quantity of radioactive substances, which remain active for long periods of time and reach plants and ultimately animals through fallout and in rain and snow. The exploitation of atomic energy for peaceful purposes also results in radioactive waste, the removal and safe storage of which are still great problems, quite apart from the possibility of their liberation by accident. It has been established that all over the earth Man is exposed to a considerably higher degree of radiation than was the case before the time of atomic explosions, and it is obvious that this will get worse and worse if human reason does not prevail. It is known that germ cells can be influenced by the smallest doses, in proportion to the applied radiation, and that the effects of exposure are cumulative. This means that the germ cells, which cannot "forgive and forget," hand on the sum of our sins against them from one generation to the next. They live on, while the individual dies. As time goes on a great increase in mutations may take place, and we know that most of them will be unfavorable. This ever-present danger now threatens the life of future generations, although years or even several decades may pass between the first exposure to radiation and the appearance of the final dire consequences in the human race.

Man is aware of the danger, but it is questionable whether he has the sense to keep it in check. Incredibly careless about the welfare of future generations, he plays with fire before he has learned how to control it.

Let us summarize: Modern genetics has clarified the meaning of the term "variability." There are on the one hand changes in the external appearance of an organism, brought about by the influence of the environment, that are not heritable and that are called modifications. Quite different on the other hand are those changes in the external appearance that are due to heritable alterations of the genes within the germ cells, known as mutations. The frequent occurrence of natural mutations had at first escaped the notice of geneticists. Now they can increase the number artificially by special manipulation. Thus the existence of mutations is one of the most powerful arguments in favor of Darwin's theory, however much it seemed initially to contradict it. In studying Darwin's work, we find that he was quite aware of the existence of heritable and nonheritable variations and that he believed in the possibility of the inheritance of so-called acquired characteristics, provided the external causes get the chance to exert their influence for a long enough time to work the change.

4. Can Acquired Characteristics Be Inherited?

What a silly question, you may think. Every characteristic has once had to be acquired, and since many characteristics are heritable, acquired characteristics must be handed on. But this is not what we are discussing. The question does not refer to mutations, which are changes of factors in the germ cells and are therefore heritable. This much and heatedly discussed question refers to characteristics that are acquired by an actively working organ and affect the body rather than the germ cells.

For quite a long time now it has been the custom to cut the tail of certain breeds of dog, in order to improve their looks, but no sensible person would really expect them one day to produce puppies with short tails.

In fact, there is a vast amount of evidence to show that the effect of the environment on the body cells is not passed on to the germ cells (the eggs or sperms). August Weismann cut off the tails of mice generation after generation only to find the young born with tails — as he expected. He concluded that to affect a physical character it was the *germ plasm* (the egg-producing or sperm-producing cells) that had to be changed — not the body cells, the somatoplasm. It was the germ plasm that provided continuity, not the somatoplasm. Or as we have said, the chicken is but an egg's way of producing an egg. Or still another way, the somatoplasm (the body) is the germ plasm's way of producing more germ plasm. The germ plasm is continuous throughout the ages. New characteristics are acquired only if the germ plasm is changed — and the change is passed on.

The effect of use and disuse

Lamarck proclaimed the great importance of the inheritance of acquired characteristics in the formation of new species and as the basis for evolutionary adaptation. It is a fact that the more frequently and vigorously parts of the body are used during the lifetime of an individual, the more strongly they develop. The muscles of an athlete are better developed than those of a bookworm. But the children of the athlete are not born with the athlete's muscles; they need to develop them through exercise. If a patient's diseased kidney is removed the remaining one has to do more work, and consequently it increases considerably in size. But the children of such a person are born with two kidneys; it is the germ plasm that is passed on.

On the other hand the tailed amphibian Proteus, which lives all its life in the eternally dark caves of the Dalmation Karst and never makes use of visual organs, has very tiny eyes which are grown over by skin. Certain snakes have just remnants of limbs, which their ancestors once used freely, and the modern horse still has the remnants of the originally five-toed foot. Thus we know quite a number of vestigial organs in various stages of reduction until they finally disappear completely. These phenomena would find a convincing explanation if it could be shown that the effects of disuse are heritable and can induce progressive structural and functional reduction of an organ.

Doubts

When Darwin set out to explain the origin of species, he accepted Lamarck's theory of the inheritance of acquired characteristics as a valid addition to his theory of natural selection. Modern geneticists reject Lamarck's theory for two reasons. Despite all efforts the inheritance of acquired characters has never been demonstrated by experiment. Furthermore, how can an acquired character, say a muscle, grown strong through use, possibly influence the far-removed germ cells of the individual in question, so that the very muscle develops more strongly in its offspring?

The failure to demonstrate experimentally the inheritance of acquired characteristics could of course be due to an insufficient time of exposure to whatever causes the change. What are a few years of experimenting compared with the epochs of time during which the formation of species took place? Nevertheless things happen whether we comprehend them or not. We know for certain that the body and mind of a person, posture and features, and mental powers are inherited through the microscopically small sperm and egg cells.

Is this really comprehensible?

Hereditary callosities

There are adaptations of which it is difficult to believe that they are due to mutation and are the end result of selection. They forcefully suggest the possibility of an inheritance of acquired character. The inheritance of functionally acquired callosities of the skin seems to me such a case.

Everybody knows that when we row a lot, the oars produce calluses in the palms of our hands. Much-used parts of the skin grow thicker. This is one of the many purposeful reactions of organs. The skin on the sole of our foot is always more exposed to mechanical wear and tear than the top of the foot and is accordingly tougher. In the human embryo, that part of the skin of the foot that it will later walk on already begins to thicken. This cannot be a process of natural selection, as the callosity stops exactly at the rim of the sole and never grows beyond it. It could not mean a handicap in the struggle for existence if the callosity grew a little above the rim of the sole. The thickening of the sole, covering exactly the region of future use in the yet unborn embryo and representing the region of former use by countless previous generations, eludes our understanding if we are not prepared to assume this to be a functionally acquired characteristic.

A similar case is represented by the African wart hog, which has the strange habit of crawling on its knees while digging for food. Anatomically speaking, the so-called knees are its wrists. The skin on the back of the wrists has callosities and, like the soles of our feet, this region becomes thickened before the animal is born. Yet another case of a heritable callosity is shown by a bird. The hoatzin lives in the marshy parts of the Amazon River and has a crop of fantastic size, which is formed by an impressively enlarged and convoluted esophagus. Within it the hard leaves of its favorite food plant are stored and prepared for further digestion. The filled crop makes the bird top-heavy and if, after a cumbersome flight, it alights on a tree, it always has to rest its chest on the branch it has settled on. Exactly above that part of the breastbone that supports the front part of the bird's body is found a callosity that is already laid down in the young birds before hatching.

Even to discuss the possibility of inheriting acquired characteristics will arouse the horror of most geneticists. In numerous sets of experiments carried on over long stretches of time, they have discovered conclusively that environmental influences are not inheritable. They are convinced that this pragmatic test applies also to the much longer periods in the evolution of the species. But the most protracted experiments of modern genetics cover only a few years, a brief span indeed compared to the experiments of Nature, which reckons in millennia. The question is: Will the concepts acquired by one generation of geneticists stand the test against the experience of those vast periods of time?

But it may well be that the science of genetics will someday throw new light into the places where unknown forces still work in darkness.

5. *The Success of Planned Breeding*

It is quite understandable that geneticists are enthralled with the theory of natural selection and discard the theory of the inheritance of acquired characteristics. After all, the many and latest successes in practical breeding are based on planned selection. Never yet have changes in the environment led to a change in hereditary factors or to an improvement of a variety.

Artificial selection has been practiced since time immemorial, and it has produced the different breeds of our domestic animals. To Darwin it was the decisive stimulus to put forward his theory of natural selection. The method of the earlier breeder was, however, very much one of trial and error. Our newly acquired insight into the correlations between hereditary factors and variability has made it possible to work out more ambitious plans for breeding bigger and better specimens.

Peculiarities produced by cultivation

Domestic animals and cultivated plants are all descended from wild stock. From this they differ by showing all those characteristics that Man desires them to have. The pig of today can grow an amount of fat the wild pig could never produce, and that even a domestic pig of several hundred years ago could not muster. The now cultivated races of cereals produce grains in such numbers and of such size that their ears look quite different from those of the wild-growing species from which they are derived.

Cultivated forms differ in yet another point from the wild stock. The latter are uniform, while the former show extraordinary variability. We need only remind ourselves how much domesticated dogs vary in color and pattern of their hair coat, in size, in watchfulness and temperament, and how similar to one another their nearest wild relatives, the wolves, are. There are two reasons for the variability of cultivated forms, and one had better be clear about them if one wants to be a successful breeder.

First of all: just as in the untended natural state so among domesticated animals and plants heritable mutations may suddenly appear. In Nature these would be eliminated in the shortest of time in the struggle for existence. Under the protective hand of Man, even freaks that are not well equipped for life can and do thrive. Often it is just these that are specially cared for. In Lapland it has been observed that the white offspring of reindeer are nursed with special care by women and children. They love to breed these freak animals which in Nature would not have a chance of surviving. Thus this particular mutation is fostered and the constant backbreeding with the wild form is the source of an increase in variability of the cultivated form.

The second reason for increased variability lies in the fact that breeders very often purposely crossbreed with more distantly related varieties in order to add a few desirable features. The two varieties often differ in quite a number of heritable factors. When we discussed Mendel's laws we learned that a cross between two varieties, differing in more than one characteristic, necessarily leads to greater variability and to new assortments of hereditary factors. If two varieties differ in ten characteristics showing Mendelian inheritance, we get 1,024 different possible combinations. With twenty different characteristics more than a million different offspring will show these characteristics in all possible permutations, from which one can choose to breed. In such a pool whatever wanted can be found. For centuries now breeders have practiced this kind of selection, guided by their experience. Only relatively recently have they started to benefit from the scientific insight into the laws of inheritance and variability.

Science the handmaid of agriculture

Vineyards the world over were nearly brought to complete destruction by two very fierce enemies, the grape phylloxera and mildew, among other fungal diseases. Both can be controlled by spraying, but the method is expensive and tedious. There are wild vines in America that are resistant against these pests, but they produce inferior quality grapes. Both varieties (the wild and cultivated) can be crossed with a view to obtaining good grapes as well as pest-resistant plants. The plant-breeding station of Müncheberg, which was founded by Erwin Baur, produces about 7,000,000 seedlings yearly, from such crossbreedings. All of these were experimentally inoculated with mildew, and the 199 out of 200 that succumbed were destroyed. From the remainder those prone to grape phylloxera and those showing an unsatisfactory yield were then eliminated. After several decades of experimentation, carried out in Germany, France, and America, since 1947 promising strains have been produced, which are now being checked for their suitability to different soils and climates. It is hoped thus to collect from millions of combinations a vine that answers all requirements.

In a similar way strains of tomatoes have been crossbred and selected from, to obtain plants that ripen a few weeks earlier. This saves great sums of money on imports of early tomatoes from warmer regions.

In the eighteenth century sugar cane was still the only known source of sugar. The discovery by the chemist Andreas Marggraf that beets contain the same kind of sugar was at first without practical significance, because it was not worth while to try to extract the 6 per cent of sugar from beets. By 1910 selective breeding had raised the sugar content to 18 per cent, and now we have strains of beet containing up to 26 per cent of sugar.

Probably the most spectacular success was achieved by Baur and his coworkers with lupines. The ordinary lupine can be used only to improve the soil; in its root live nitrogen-fixing bacteria, and by planting lupines the nitrogen content of the soil is increased. The plant as such and its fruit are bitter and quite useless as fodder even for grazing animals.

As a result of selective breeding of 1,500,000 plants, a very few were obtained that had a negligible content of bitter substances. These "sweet lupines" bred true. They are a great gain to agriculture and are a perfect substitute for expensive protein fodder, which hitherto had to be imported in large quantities.

These are just a few examples out of many. One may deserve special mention. It concerns the work carried out in Sweden. There they crossed their own winter wheat with English wheat, which, though of greater yield, does not stand the rough Swedish climate. Finally a strain of wheat emerged that combines rich yield with resistance to cold. In consequence Sweden today produces 70 per cent more wheat per acre than fifty years ago.

Few of us who profit by such progress are aware that we are reaping the fruits of Gregor Mendel's and his successors' labors. The work of these men, which was done for purely theoretical reasons, has become the foundation of all the planned breeding that is undertaken today at great practical profit. But scientific progress moves swiftly, and every country that does not want to be left behind has to keep up with this accelerated pace.

6. *Man, His Past and His Future*

In the zoo

Zoos exhibit a number of apes that are called anthropoid, or manlike. Among them are the chimpanzee, the orangutan, and the gorilla. In the zoos of earlier days they sat about apathetically. Now they are given something to do and are frequently taken for a walk within the park. On such an occasion a young chimpanzee is apt to jump onto the back of its mother and want to be carried. If the keeper interferes, pulls it down, and tries to lead it by the hand, the little one throws itself on the ground, screams and has tantrums, gets up and runs into the shrubbery, where in a fit of anger it tears leaves and branches to pieces. The striking similarity to the behavior of a naughty child raises shrieks of delight among the onlookers. Such scenes are most impressive. In fact every time we watch apes in their cages we are startled by the manlike expression and behavior of these animals. The monkey house exerts a strange fascination. The visitors would be even more startled if they were fully aware of all the existing similarities between them and these animals. This is not restricted to external behavior only; it includes all the organs, the whole skeleton, every single bone and tooth. The brain of a chimpanzee has the same internal structure and on its surface the same pattern of furrows and folds as the human brain, which, however, is three times as large. The mother chimpanzee nurses its young and its placenta is not unlike that of the human mother. These and thousands of other features point to an intrinsic relationship, extending to a blood relationship in the truest sense of the word. This can be visibly demonstrated by chemical methods.

The test for blood relationship

The biologists have worked out an excellent method to prove blood relationship:

If a few drops of the blood of two different animals, for instance of a rabbit and a dog, are mixed together in a little glass dish nothing happens. If a considerable quantity of blood of a dog is injected into a rabbit it gets cramps and dies, because the non-related red blood cells clot and the blood itself produces unfavorable reactions. If the injections are started in very small doses and then repeated in slightly increasing quantities, the rabbit develops antibodies against the foreign blood. If this blood is then mixed in a glass dish with the blood of a dog little flakes of protein are precipitated. This is similar to the formation of antibodies against the toxin of bacteria. The antibodies of the rabbit, the formation of which are stimulated by the injections of the dog's blood, now coagulate the protein in the dog's blood so that it can be seen in the form of flakes.

As with the antibodies against bacteria, this reaction is a specific one, which means that if we inject dog blood into the rabbit the antibodies formed will be effective only against dog's blood. If we add the blood of a wolf, which is a relative of the dog, we get a slightly weaker precipitation; if we add the blood of a nonrelated animal, or Man, we get no precipitate at all. If, however, human blood had been injected several times into a rabbit, the rabbit's blood will precipitate human protein. But now comes the important point: if the blood of such a rabbit is mixed with the blood of a chimpanzee a precipitation of protein occurs, which is only slightly weaker than that caused by human blood.

This reaction is so reliable and succeeds with such small quantities, even of dried blood, that it is used in court as evidence to distinguish between animal and human blood. A rabbit that has been injected with the blood of a chimpanzee reacts most strongly against chimpanzee blood, less strongly against human blood, and not at all against the blood of monkeys. This is a new and ingenious way of confirming a result that had long before been worked out from anatomical studies, namely, that the blood relationship between ape and Man is much closer than between ape and monkey.

Prehistoric Man

These connections have become still more obvious through prehistoric finds. Remnants of human bones from the Stone Age differ hardly or not at all from those of present-day Man. In 1856 a human skull was discovered in the glacial deposits of a cave in the Neanderthal, near Düsseldorf. It had a beetled brow,

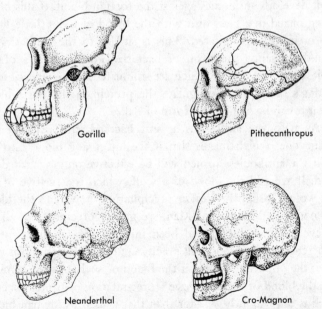

Fig. 101. Reconstruction of the skulls of early Man. The gorilla skull is included for comparison but it is not to be considered in the line of Man's ancestry.

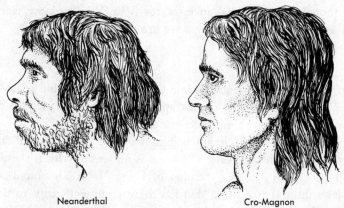

Neanderthal   Cro-Magnon

Fig. 102. A reconstruction of the heads of Neanderthal and Cro-Magnon Man. Their clear relation to modern Man is unmistakable.

a receding forehead, a large lower jaw without chin, and a small cranial vault. All these are primitive and apelike features (Fig. 101). Other finds of bones complete for us the picture of the Neanderthal Man, and we have a good idea of what he may have looked like. From his cave dwellings we know that he used primitive tools made from stone. The Neanderthal Man was not, as was assumed for a time, a direct ancestor of modern Man but a side branch on the human genealogical tree. Immigrants from the East brought about his extinction.

In still older geological layers in Java the now famous Pithecanthropus, or ape-man, was discovered. Related finds are known from Africa and Europe.

The modern gorilla is definitely not our immediate ancestor (Fig. 101). We do not even know for certain whether modern Man has descended from Pithecanthropus or from one of the related side branches. The finds demonstrate, however, without doubt, that the gap between the most highly evolved apes and prehistoric Man was then still narrower than it is today. Reconstructions (Fig. 102) from the skulls of Neanderthal Man and Cro-Magnon show their clear relation to modern Man *(Homo sapiens)*.

I have never quite understood why people take exception to

the thought of an animal ancestry for Man. Whatever we were in the past cannot change what we are today.

The future

Our past has left its traces, while the future of mankind is wide open to speculation. We cannot know it, we can only imagine it. And as we do so we find grave reasons for concern.

The evolution of living beings has been going on on this earth for millions of years, and there is no reason why it should stop now. Man too, as a member of living Nature, is bound by its laws. Like the animals, Man has to adapt himself daily to the demanding tasks of life, and in the struggle for comfortable living conditions for himself and his family, the better man is the enemy of the good one.

In the very essential field of physical and mental health civilization has blunted the fierceness of selection. Common humanity and medical skill foster the survival of abnormalities that would be mercilessly eliminated among primitive peoples or among wild animals. The lame and the blind are fed as everybody else is. Weak children are reared with all the means at our disposal, born cripples are carefully nursed, the mentally deficient are looked after by the state, with money earned by the efficient citizen. The feeble-minded are allowed to reproduce; in fact, they do so often without any inhibitions or sense of responsibility. In one case, the existence of seventy-five living feeble-minded persons could be traced back to a single individual with heritable feeble-mindedness.

This cannot be called an efficient selection. In fact, inferior traits are favored not only by letting idiots breed, but by the widespread trend among the economically secure to keep the number of their children down by birth control, as also by the way in which we fight wars, exposing the mentally and physically fittest to the greatest dangers. All this may quite obviously lead to a deterioration of the human race.

Eugenics

To prevent such fatal consequences is the aim of eugenics. Just as medical science benefits the health of the individual, eugenics is concerned with the physical and mental welfare of coming generations. The basis for a healthy population lies in its healthy hereditary traits.

The people who populate this earth differ considerably from one another. Geographical isolation, differences of climate, and quite a few other circumstances unknown to us have led to the development of races that differ to a varying degree in their physical characteristics and in their mental equipment. In earlier days mountains, deserts, and the sea were efficient barriers; nowadays the development of modern means of transportation leads to frequent and ever-increasing contact. The races are brought face to face, they work together and begin to mix, and genetical hybridization occurs between members of races that differ considerably in their hereditary make-up. We know that the ensuing progeny shows a much-increased variability. The animal or plant breeder uses this knowledge when he hybridizes different races and chooses, from the wide range of new combinations, those few that suit his particular purpose. Nobody, however, has the right or the power to suppress, like the breeder, the undesirable human types that may result from hybridization. Therein lies a certain danger that an intelligent application of eugenics could minimize.

Hereditary material could be improved if people with heritable diseases were prevented from reproducing. To exclude dangerous hereditary factors in this way would mean nothing less than to replace the dwindling influence of a natural selection by a vigorous artificial selection.

In ancient history we come across attempts to prevent heritable disease by a kind of artificial selection. But to return to the practice of the Spartans, who carried weak children into the mountains and let them starve to death, would not only go against our modern way of feeling, but would serve no purpose because weak boys often grow into strong men. What counts are the hereditary factors, not the external appearance or a passing phase during a certain period of life. Unfavorable environmental conditions such as might prevail during the development of the embryo can certainly affect the best hereditary factors, and the result will be a weak baby. This leads us to the following question: So far as Man is concerned do we know with any certainty which of the characteristics of an individual are heritable? It is relatively easy to experiment on animals and collect information. With human beings this is much more difficult because one cannot experiment on them. Nature herself, however, obliges us with one kind of experiment which even the most original research worker could not have figured out more ingeniously: she offers us the identical and other twins for observation. We have talked about them before. One example may illustrate how they tie up with our question.

A short visit to the operating theater

If, during a difficult operation, the life of the patient seems in danger, the surgeon may have to use a blood transfusion. A healthy person donates a certain amount of blood, which is gently run into the blood vessels of the patient. In the early days of this practice the desired effect frequently not only failed to come about, but additional complications set in. The reason lies in differences in the composition of blood. According to certain properties of their blood human beings can be divided into four major blood groups. Some of them are compatible with each other and others are not. It is easy to find out to what blood group an individual belongs, and we can choose a suitable donor for every patient.

However, all this raises a question: What makes us belong to a certain blood group? Is it a matter of heredity, or is it decided by environmental factors during our development, or are both working together and equally important? How do we establish whether and to what degree a human characteristic is heritable? Here, once more, the study of twins helps us to clear up doubts. Identical twins are derived from one and the same fertilized egg, therefore they possess identical hereditary material while other twins differ from one another as much as any ordinary siblings. It has been proved that identical twins invariably belong to the same blood group, while other twins do not. From this it follows that the type of blood group of an individual is determined by heritable factors and cannot be influenced by the environment.

The comparison between identical and other twins has helped to clear up still more involved problems about the relative importance of heredity and environment.

Heritable diseases

Let us concern ourselves a little longer with our blood. Normal blood clots. As soon as it leaves an open blood vessel it coagulates, closes the wound, and thus prevents excessive bleeding. In a relatively rare disease, hemophilia, blood fails to clot, and a hemophiliac can bleed to death from a small wound. This disease is heritable, and the somewhat involved way in which it is handed on is now exactly known. Until recently most bleeders died before they had a chance to marry. Up to now, therefore, the disease had no chance to spread unduly.

But means have been found to stop the bleeding and the patients need no longer bleed to death. They will live to a greater age and will have a chance to hand on their defective hereditary factor to the coming generation. Thus the disease will spread and the doctor, the ministering angel to the suffering individual, contributes at the same time toward the deterioration of the race.

What the future will bring is already history as far as some other hereditary shortcomings are concerned. A quarter of our grown-up population is nearsighted. With our ancestors, nearsightedness was surely as rare as it still is among primitive peoples. In competition with his normal-sighted companions, the nearsighted hunter and nomad is at a great disadvantage in the struggle for existence. The living conditions of civilized peoples, however, are as suitable for the nearsighted as for anybody else. Natural selection has no chance to stop the spreading of this condition.

Even more obvious is the case of our teeth. The skulls of Stone Age Man bear silent and yet eloquent witness that no dentist was needed then. Life was hard and kept teeth healthy. The ones who could not chew the hard and tough food, or use their teeth as tool or weapon, were no match for life in those days. In our present civilization children already need the dentist's attention. The diminished resistance of teeth is controlled by a heritable factor. In identical twins corresponding teeth begin to decay at the same time, which is as convincing as it is spectacular. Both the skill of our dentists and our cooked food result in this outcome: bad teeth are not dangerous to the survival of either the individual or the race. Consequently this shortcoming too will spread.

It would be nonsensical to try to wipe out all hereditary shortcomings in a civilized population. This would be equivalent to the extermination of the whole lot. If a pair of spectacles can turn a nearsighted person into a valuable member of the community, and if a visit to the dentist's keeps up the efficient working of a defective set of teeth, no one will mind such imperfections. If, however, a hereditary disease is disabling to such a degree that the unfortunate individual is a danger and burden to himself and the community, then the problem is a serious one.

Animals follow their urges without restraint, and Nature selects healthy and strong ones ruthlessly but efficiently. Man calls himself proudly the master of creation and yet is in danger of coming to harm. He will have to adjust his laws to his newly acquired insight, because the laws of Nature will not make any exception in his case.

Lenient measures, such as marriage guidance or prohibition to marry, are of little avail, since the urge to love and procreate cannot be kept in check by them. It can only be hoped that through sensible education of the young the vital importance of hereditary processes will become common knowledge and lead to voluntary co-operation. Perhaps, also, reason will one day prevail so that nuclear bombs will be banned from our planet, to avoid additional danger to genetic health. What is called for is a sense of responsibility, joined to a more enlightened awareness of the importance of mental and physical health, to make us more careful of our inheritance.

But what is mainly to be desired is the progressive evolution of our ethical qualities, the chief prerequisite for a better future. When people will have the insight voluntarily to avoid everything contributing to the deterioration of the future human stock, human development may one day reach the stage of relinquishing willingly all weapons of war and of sharing honestly and honorably the riches of this earth.

INDEX